Engineering Applications of Computational Methods

Volume 1

Series Editors

Liang Gao, State Key Laboratory of Digital Manufacturing Equipment and Technology, Huazhong University of Science and Technology, Wuhan, Hubei, China

Akhil Garg, School of Mechanical Science and Engineering, Huazhong University of Science and Technology, Wuhan, Hubei, China

The book series Engineering Applications of Computational Methods addresses the numerous applications of mathematical theory and latest computational or numerical methods in various fields of engineering. It emphasizes the practical application of these methods, with possible aspects in programming. New and developing computational methods using big data, machine learning and AI are discussed in this book series, and could be applied to engineering fields, such as manufacturing, industrial engineering, control engineering, civil engineering, energy engineering and material engineering.

The book series Engineering Applications of Computational Methods aims to introduce important computational methods adopted in different engineering projects to researchers and engineers. The individual book volumes in the series are thematic. The goal of each volume is to give readers a comprehensive overview of how the computational methods in a certain engineering area can be used. As a collection, the series provides valuable resources to a wide audience in academia, the engineering research community, industry and anyone else who are looking to expand their knowledge of computational methods.

More information about this series at http://www.springer.com/series/16380

S. Arungalai Vendan · Rajeev Kamal ·
Abhinav Karan · Liang Gao ·
Xiaodong Niu · Akhil Garg

Welding and Cutting Case Studies with Supervised Machine Learning

Springer

S. Arungalai Vendan
Department of Electronics and
Communication, School of Engineering
Dayananada Sagar University
Bangalore, India

Abhinav Karan
School of Engineering
Dayananada Sagar University
Bangalore, India

Xiaodong Niu
Shantou Ruixiang Mould Co. Ltd.,
Jinping S&T Park
Shantou, China

Department of Mechatronics Engineering
Shantou University
Shantou, China

Rajeev Kamal
School of Engineering
Dayananada Sagar University
Bangalore, India

Liang Gao
State Key Lab of Digital Manufacturing
Equipment and Technology
Huazhong University of Science
and Technology
Wuhan, Hubei, China

Akhil Garg
School of Mechanical Science
and Engineering
Huazhong University of Science
and Technology
Wuhan, Hubei, China

ISSN 2662-3366 ISSN 2662-3374 (electronic)
Engineering Applications of Computational Methods
ISBN 978-981-13-9384-6 ISBN 978-981-13-9382-2 (eBook)
https://doi.org/10.1007/978-981-13-9382-2

This Springer imprint is published by the registered company Springer Nature Singapore Pte Ltd.
The registered company address is: 152 Beach Road, #21-01/04 Gateway East, Singapore 189721, Singapore

Preface

This book presents machine learning concepts applied in engineering mathematics for applications in advanced welding and cutting processes. Few welding and cutting case studies are presented with details on experimentation and characterization. Subsequently, parametrical interdependencies of various entities governing the processes are investigated using data analysis and data visualization techniques are an embodiment of machine learning. The contents present fundamental and advanced terminologies of supervised learning where focus is laid on Python libraries such as NumPy, Pandas and scikit-learn programming. It emphasizes on the features and benefits of employing machine learning techniques for quantitative analysis of manufacturing processes in the engineering domain. The book exposes the beginners to basics of machine learning for applied sciences, enabling them to acquire requisite knowledge on data sets and its branches for information excavations and adapt to the global competitiveness and work on real-time technical challenges of data. Besides, it also acts as a valuable resource for scholars with ample domain knowledge using conventional mathematical tools for data analysis.

Bangalore, India — S. Arungalai Vendan
Bangalore, India — Rajeev Kamal
Bangalore, India — Abhinav Karan
Wuhan, China — Liang Gao
Shantou, China — Xiaodong Niu
Wuhan, China — Akhil Garg

Contents

Chapter 1
Supervised Machine Learning in Magnetically Impelled ARC BUTT Welding (MIAB)

1.1 Introduction

The MIAB welding process was initially investigated by the E. O. Paton Electric Welding Institute during the 1950s. It was later developed for commercial applications by Kuka Welding systems, who named it the Magnet arc process. Today, MIAB welding is used for a variety of applications throughout Europe and Ukraine [1].

Magnetically impelled arc butt (MIAB) welding is a unique and advanced process which utilizes relatively simple equipment, though based on a set of complex interactions among an arc, an applied magnetic field and also an induced magnetic field. These interactions are accompanied by various changes in terms of arc length, temperature distribution, electromagnetic flux distribution, electrical and mechanical properties of the material, etc. which occur during the heating of the parts being welded. The result is a swift welding process that offers cost savings for a range of joint configurations.

This process is generally employed for thin-walled tubes up to 4 mm in the automobile, machine building, construction and other processing industries. MIAB welding is extensively used in automobile industries in European countries and seldom used in parts of the United States and the United Kingdom.

1.2 Process Principle

MIAB welding is a fully automated solid-state welding process. An electric arc is made to strike between two tubes which are aligned in axial direction with a small gap in between. The arc is impelled to move around the joint line by the force of interaction between the arc current and an externally applied magnetic field.

S. A. Vendan et al., *Welding and Cutting Case Studies with Supervised Machine Learning*, Engineering Applications of Computational Methods 1,
https://doi.org/10.1007/978-981-13-9382-2_1

The radial component of the magnetic flux density B_r and the axial component of the welding arc current I_a interact with each other exerting a force on the arc [2]. The mathematical expression of this electromagnetic force is given in Eq. (1.1). This force impels the arc along the peripheral edges of the tubes.

$$\vec{F}_{B_r} = K \cdot \vec{I}_a \times \vec{B}_r \tag{1.1}$$

Coefficient K depends on the value of the arc gap between the two tubes to be welded.

The force exerted on the arc current influences the speed of the rotating arc. Therefore, it is clear that adjusting the strength of the magnetic field, the magnitude of the arc current, or the width of the arc by changing arc current plays an important role on the speed of arc. In particular, by sharply increasing the welding current for a short time just prior to upset, a rapid expulsion of molten metal occurs which enables cleaning action. This eliminates the need for shielding gas.

The rotating arc heats up the peripheral edges of the tubes to cause localized melting and adjacent softening in the heat-affected zone (HAZ) and consequently lowers the yield strength of the adjacent solid material to permit sufficient forging action, a critical aspect of the process. The forging expels most of the molten metal present and a solid phase bond is formed. A basic schematic of MIAB welding is shown in Fig. 1.1, which depicts the welding of two tubes.

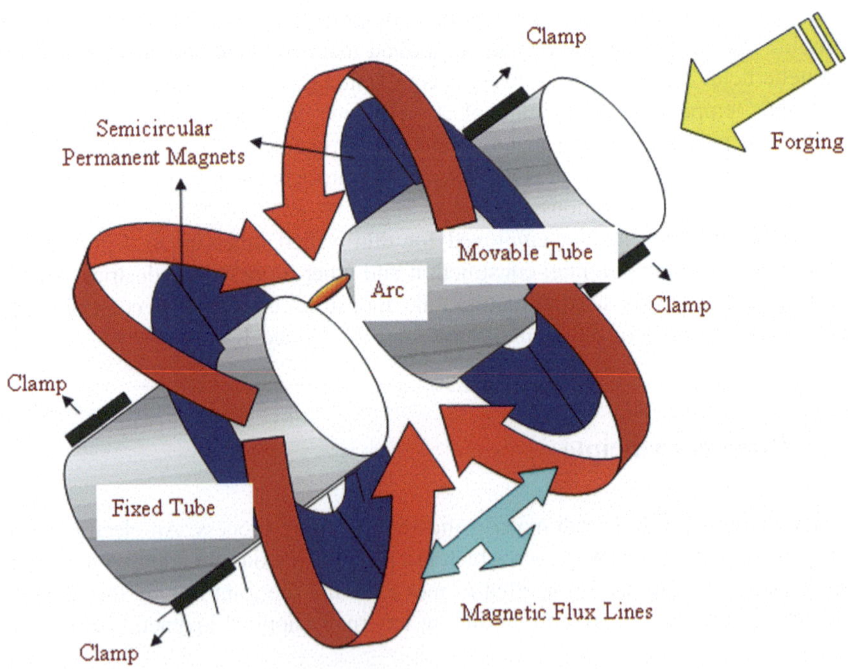

Fig. 1.1 Basic representation of the MIAB welding process

As Fig. 1.1 indicates, an arc is impelled along the peripheral edges of the tubes due to a magnetic field set-up using either permanent magnets or electromagnets. The linear speed of the arc is approximately 250 m/s. The swiftly rotating arc, in combination with the thermal conductivity of the tubes being welded, causes uniform heating at the joint. Subsequently, the molten abutting faces are pressed together by a forging cylinder. This upsetting operation expels the molten material out of the joint and creates a forging action on the remaining plasticized metal. This forging action then forms a porosity-free weld. The process does not use filler material. Shielding gas is usually not required [4]. When shielding gas is not used, as in the case of this research, a short pulse of high current is included which expels contaminated molten metal prior to upset.

MIAB welding is carried out in two stages:

A. **First stage**

The first stage of MIAB welding involves co-axial alignment and clamping of the two tubes with a small gap between them. Then a DC current is made to flow through the circuit formed by the power supply, the clamps and the tubes. Further, a carbon rod is employed to strike an arc. Figure 1.3a depicts the situation wherein the arc moves along the tube edges.

The arc is swivelled along the peripheral edges of the tubes at a high speed. Arc rotation persists for a few seconds until the faying edges are heated to a high temperature, i.e. beyond softening temperature.

B. **Second stage**

After the edges reach a suitable temperature they are pressed together with a pre-determined forging pressure and the weld is set as shown in Fig. 1.3b.

The direction of the force is determined by applying Fleming's left-hand rule, according to which the rotating direction of the arc is always perpendicular to the applied magnetic field and the arc current as shown in Fig. 1.2. The force occurs due to the magnetic flux lines generated by the flowing current interacting with the magnetic flux lines of the applied magnetic field. This phenomenon is shown graphically in Fig. 1.4, which depicts a current carrying conductor under the

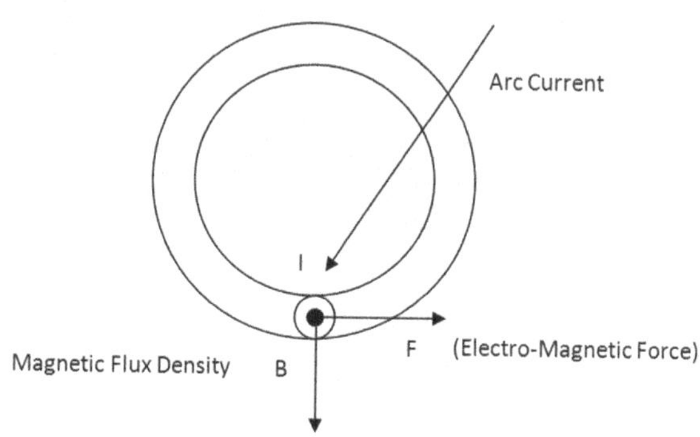

Fig. 1.2 Fleming's left-hand rule [3]

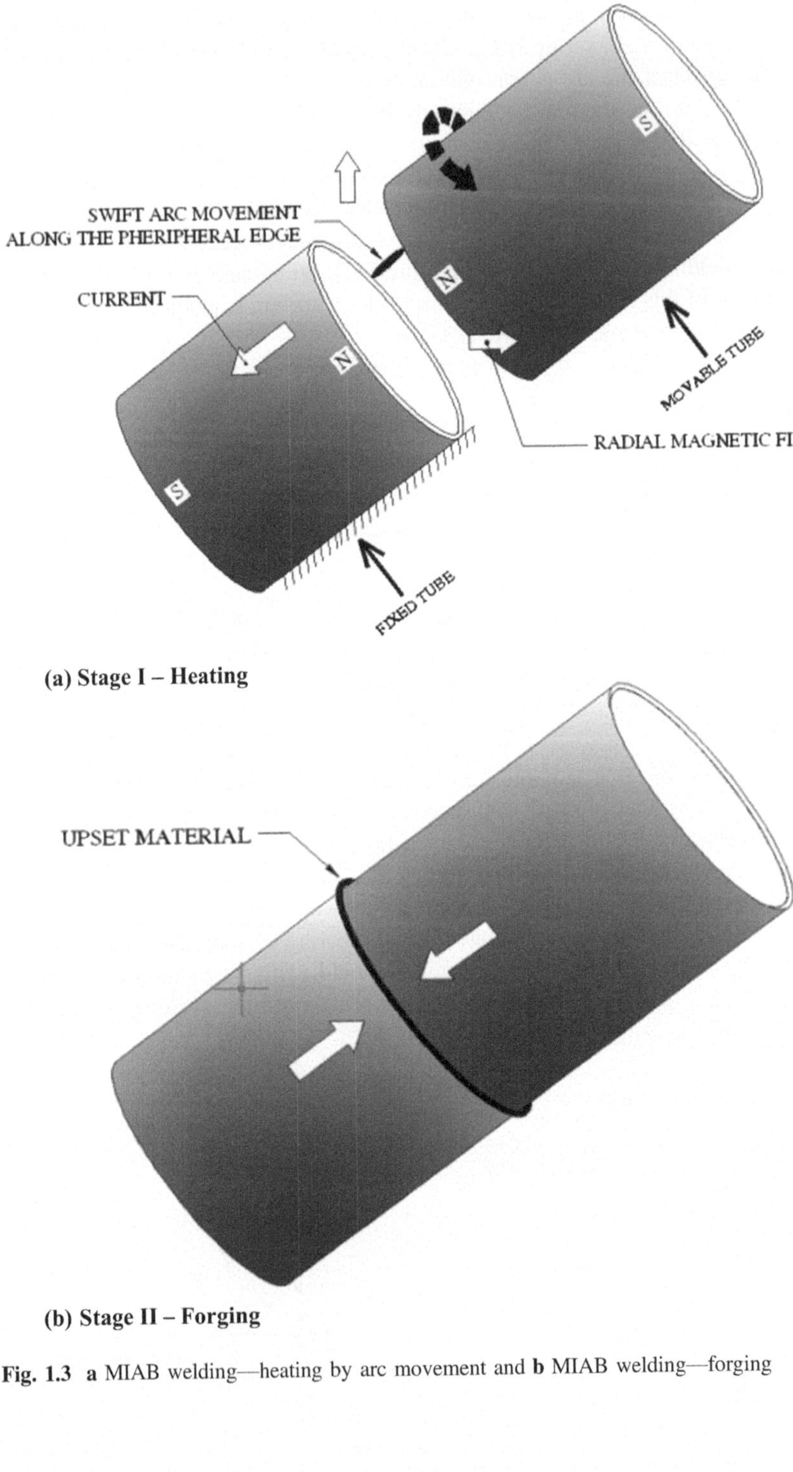

Fig. 1.3 **a** MIAB welding—heating by arc movement and **b** MIAB welding—forging

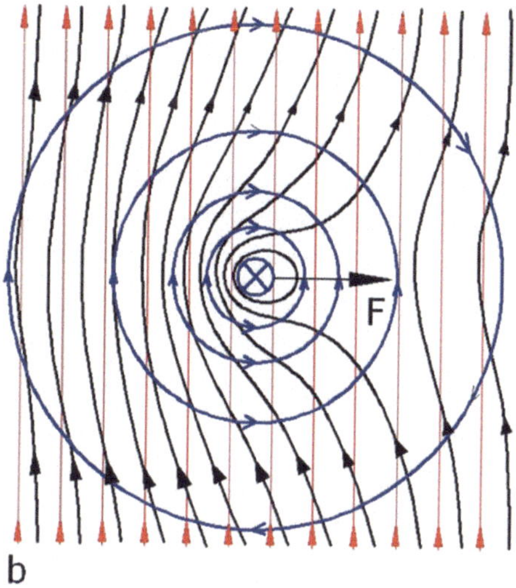

Fig. 1.4 Magnetic field exerting force on current carrying conductor [5]

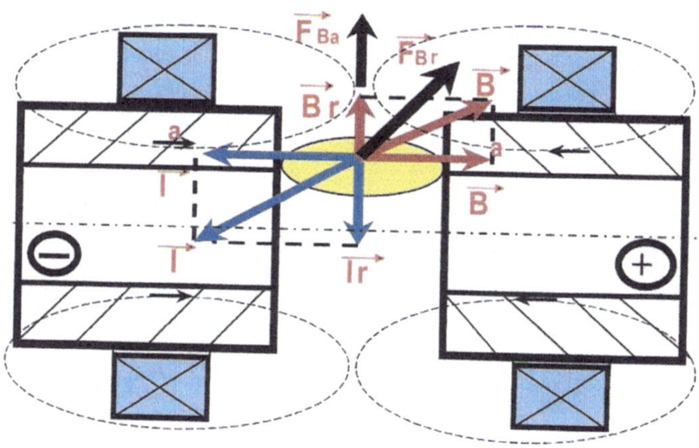

Fig. 1.5 Effect of radial component of arc current on arc movement [2]

influence of an applied magnetic field. The force is generated on the side of the conductor where the magnetic flux lines are dense [5].

In addition to the primary force on the arc that causes the arc to spin rapidly around the part, there is an additional important force on the arc. This force is generated when the radial component of the arc, I_r, crosses the axial component of the magnetic field B_a, as shown in Fig. 1.5.

Initially, while MIAB welding a ferromagnetic material, the arc is pushed to the ID of the joint due to arc blow effects. Upon heating, the Curie temperature is first reached on the OD of the tube altering the distribution of magnetic flux in the joint and pushing the arc outward. The outward movement of the arc can play an important role in generating uniform heating at the joint [2].

Applications
The range of tube diameters which can be welded on commercially available equipment is approximately 10–220 mm with wall thickness of 0.7–13 mm (7 mm and above has been MIAB welded but with a more complex weld cycle which included the orbital motion of one tube). This comprises different types of tubes and pipes

1.3 Process-State of Art

For a researcher investigating MIAB welding with a view to applying, it for welding boiler tubes, the literature available is to some extent, scarce and inadequate. Many of the papers are translated from other foreign languages. In some cases, the translations are not clear; adding to the difficulty in understanding. This chapter presents a comprehensive overview of the earlier research work carried out in the area of MIAB welding. It also reveals the disagreement among various researchers with respect to the underlying concepts of MIAB welding; in particular, the discrepancies pertain to the interaction between the arc and the applied and induced magnetic fields.

1.3.1 Preliminary Explorations on MIAB Welding Process

This section reports the initial postulates on MIAB welding process (Table 1.1).

$$T(y,t) = \frac{qy}{2\lambda\pi}\left\{\left[\frac{\sqrt{4at}}{y}\exp\left(-\frac{y^2}{4at}\right) - \sqrt{\pi}\left[1-\Phi\left(\frac{y}{\sqrt{4at}}\right)\right]\right]\right\} \tag{1.2}$$

where

T	temperature (°C)
t	arc rotation time (s)
q	heat input of the arc (cal/cm °C)
y	distance from arc along the tube (cm)
λ	coefficient of thermal conduction (cal/cm s °C)
a	coefficient of temperature conduction (cm^2/s) and
$\Phi\left(\frac{y}{\sqrt{4at}}\right)$	function of Gauss probability

Table 1.1 Brief of research reports on MIAB welding process and its technicalities

Authors and year	Experimentation	Observations	Images and equations
Steffen et al. [7]	• Investigated a variety of conditions and their effect on arc behaviour, including the use of internal and external magnets and different power sources. Steel tubes of various dimensions were used • A high speed video camera and an electronic image converter were used to study the arc	• Differences in the behaviour of the arc at the anode versus the cathode side of the joint was observed • Arc was seen to move freely on the anode side of the joint but was constricted on the cathode side • When the arc is forced to move in the presence of a magnetic field, the anode spot is blown ahead while the cathode spot trails • Arc always initiate along the inner diameter (ID) of the tube edges. This was due to the fact that the arc tends to move to the regions that heat up quickly and also to areas where the induced magnetic field surrounding the arc encounters the demagnetizing effects due to the applied field • It was postulated that as the steel melts on the ID edges, the arc length increases. The longer arc, combined with centrifugal forces pushing the arc outward, results in the arc moving toward the outer diameter (OD) as the welding progresses • Internal magnets resulted in faster starting of the weld, but external magnets provided for a more controlled rotation of the arc	NA
Nentwig et al. [8]	• Compared the effect of an internal versus external placement of magnetic coils on arc behaviour during MIAB welding of tubes	• The maximum radial flux density in the weld gap is always along the edge of the tube closest to the coil • When welding ferromagnetic materials, the magnetic flux density drops sharply along the wall thickness of the tube. This affects mainly the arc starting characteristics immediately following arc initiation	NA

(continued)

Table 1.1 (continued)

Authors and year	Experimentation	Observations	Images and equations
Kachinskiy et al. [2]	• Investigated the arc behaviour during the welding of hollow parts with wall thickness, greater than 6 mm	• It is a challenge to weld thick-walled components due to the tendency of the arc to concentrate on the ID of the component in the MIAB welding process thus resulting in uneven heating • The authors postulated that the anode and cathode spot sizes of an arc should be relatively larger than the wall thickness to achieve even heating • As shown in Fig. 1.6, arc column traces consume the ID of the thick-walled tubes during the initial stages of welding • Upon further heating, the arc column moves to the OD, but the large wall thicknesses prevent stable movement of the arc to the region of higher magnetic field induction leading to non-uniform heating • In order to improve this situation, the authors adjusted the position of the magnetic field so as to reinforce the axial component rather than the radial component of the applied magnetic field • With this modification, a larger axial magnetic flux component crosses the radial current component of the arc. This, in turn, produces a greater force on the arc, pushing it towards the OD of the tube	See Fig. 1.6
Kuchuk-Yatsenko et al. [9]	• Studied the arc behaviour during MIAB welding of a tube to a plate	• In this type of joint, the displacement of the arc from the ID to the OD was reportedly more pronounced due to the magnetic blow resulting from the interaction of the arc and the induced magnetic field	See Fig. 1.7

(continued)

Table 1.1 (continued)

Authors and year	Experimentation	Observations	Images and equations
		• This creates a greater concentration of magnetic lines of force on the ID of the tube, which pushes the arc outward as shown in Fig. 1.7. This situation can lead to uneven heating and a poor quality weld	
Sato et al. [10]	• Studied the phenomenon of the arc initiating on the ID of steel pipes and then moving to the OD, especially when welding thick cross-sections • Photo-transistors were used to assess the movement of the arc • Arc traces of the pipe ends were conducted when the arc was initiated on the OD in the presence of an applied magnetic field and in the absence of an applied magnetic field	• In both cases, the arc moved to the ID of the pipe (Fig. 1.8) • The initial movement of the arc along the ID during the initial arcing phase was not due to the arc initiating there, nor was it due to the applied magnetic field pushing it towards the ID. The authors illustrated that a magnetic arc blow effect occurs due to the tube geometry interacting with the magnetic field of the arc. This causes stronger lines of force on the OD of the tube which pushes the arc towards the ID (Fig. 1.9) • The movement of the arc from the ID to the OD at the tube was due to the variations of the magnetic field at the tube ends as the temperature rises with heating • Specifically spontaneous magnetism of iron drops with the temperature and at the Curie temperature (770 °C), iron is no longer considered to be magnetic. This creates a condition in which the magnetic arc blow pushes the arc towards the OD	See Fig. 1.8 See Fig. 1.9
Yatsenko et al. [11]	• Studied the velocity of the arc movement in the gap between a tube and a plate • Specifically, they evaluated the effect of weld parameters and the arc gap variations	• The speed of the arc depends on welding current, the magnetic field intensity, the arc gap and the temperature of the metals being welded • Scale and oxides on the faying surfaces at the joint played a role in arc velocity and mobility. The arc was also	

(continued)

Table 1.1 (continued)

Authors and year	Experimentation	Observations	Images and equations
	• Photoelectric cells and galvanometers were utilized to study the arc movement	observed to become highly mobile if scale was removed from the plate and tube ends • Distinctions between the anode (tube side) and cathode (plate side) spots were discussed • In particular the anode spot was seen to be interrupted, with evidence of jump-like movement, especially in the presence of metal vapours • The speed of the arc increased as the workpiece temperature is increased. It was suggested that this was due to the fact that the area of the anode spot increases with increasing temperature. This reduces the current density and rigidity of the arc plasma, allowing greater distortion of the arc column from the applied magnetic field. The increased distortion promotes the increase of new anode spots, allowing faster movement of the arc • A layer of molten metal soon forms a wave bridge on the tube ends which reduces the gap and consequently, the arc length. The reduced arc length causes stiffer arc plasma, which is more difficult to be moved by the applied magnetic field • This explains the drop in arc velocity after the first peak. The increase in gap then causes another jump in arc velocity. This is due to the decrease in arc stiffness described above	

(continued)

Table 1.1 (continued)

Authors and year	Experimentation	Observations	Images and equations
Taneko et al. [12]	• Studied the relationship between arc velocity, arc angle and the position at which power is supplied to the tubes • They used a voltage detector at various locations inside a carbon steel pipe, an oscilloscope and a high speed video camera to measure arc velocities and arc angles	• They concluded that due to the arc blow effect and the low electrical resistance of the tube, the current increases in the arc closer to the power supply connection on the tube. This increases the magnetic blow effect and decelerates the arc • As the arc moves away from the power supply point, it accelerates. The authors concluded that in order to support a stable moving arc, it is important to have numerous uniform contact points on the tube	
Xiancong et al. [13]	• They investigated heat flow in the MIAB weld joint	• They considered the rotating arc to be a constant heat source and applied the following heat flow equation for predicting the temperature at time t and distance y from the arc (Eq. 1.2) • It was determined that acceptable welds could be achieved at T = 1200 °C at y = 0.1 cm or at T = 900 °C at y = 0.4 cm • This formula can also be used to calculate the width of the heat-affected zone	See Eq. (1.2)
Kalev et al. [14]	• Investigated heat source (Arc)	• That a MIAB welding arc is not a constant, uniform heat source since a typical weld cycle involves different levels of current from the beginning to the end	NA

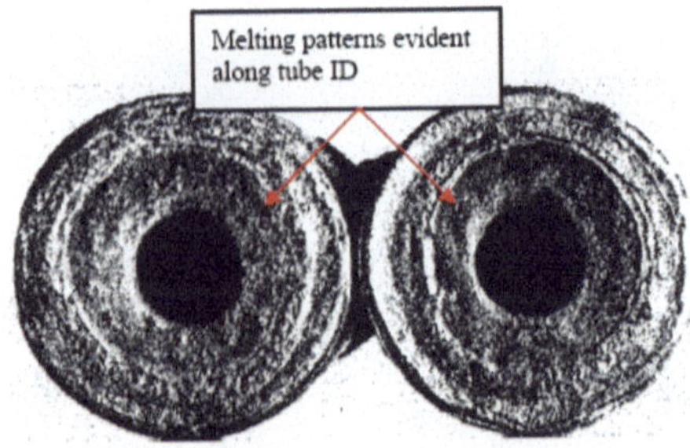

Fig. 1.6 Melting patterns of thick-walled tube reveal melting on ID

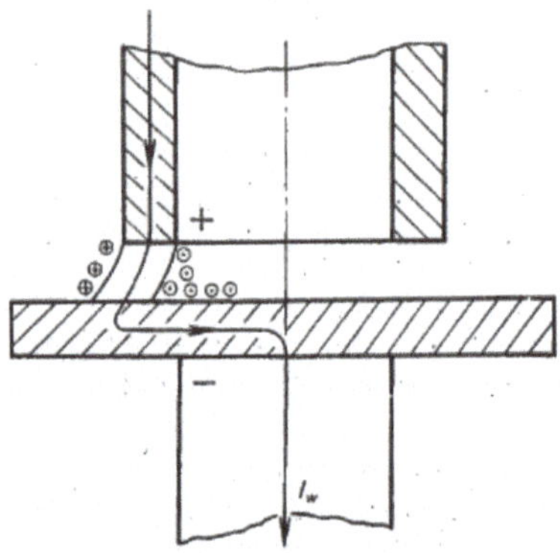

Fig. 1.7 Magnetic flux in tube-to-plate joint pushes arc outward

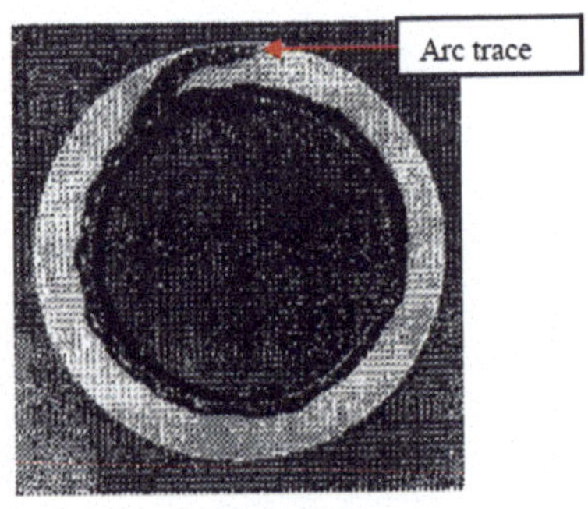

Fig. 1.8 Arc trace on tube end shows movement from OD to ID

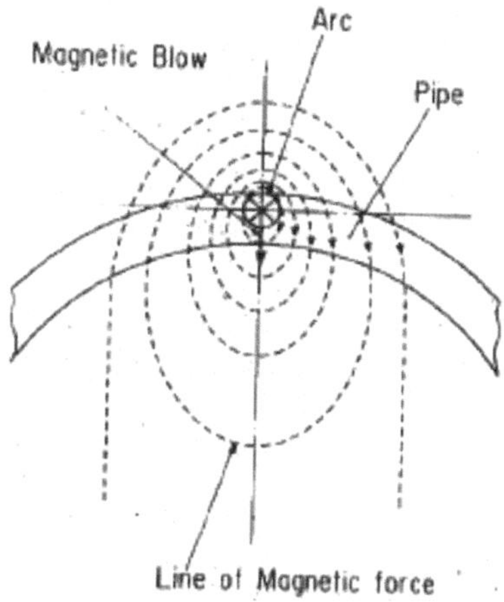

Fig. 1.9 Magnetic arc blow due to tube geometry effects

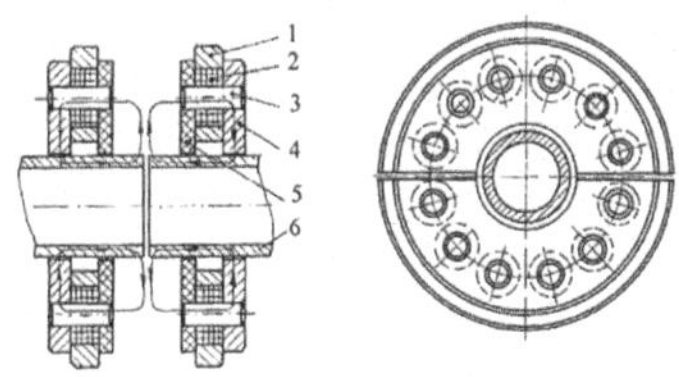

Fig. 1.10 MIAB set-up with longitudinal small coils

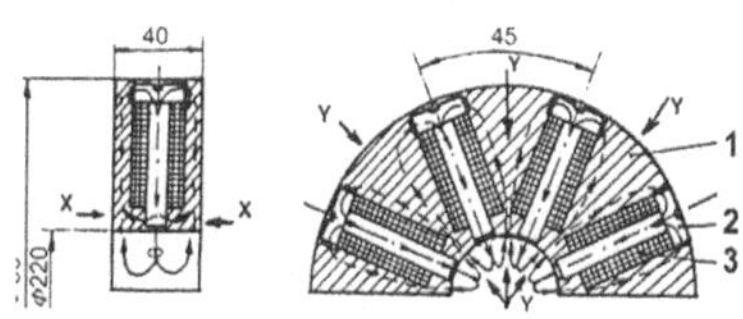

Fig. 1.11 Transverse magnetizing system

1.3.2 Design/Developmental Aspects in MIAB Welding Process

Authors and year	Experimentation	Observations	Images and equations
Georgescu et al. [15]	• Presented original ROTARC portable equipment, pneumatically operated. The equipment was designed for maximum 30 mm diameter pipes welding. Original pneumatic operating devices	• A longitudinal magnetic system design was presented that ensures easy introduction/removal of pipes because the parts were made up of two-halves, due to a system of longitudinal small coils (parallel with the	See Figs. 1.10 and 1.11

(continued)

(continued)

Authors and year	Experimentation	Observations	Images and equations
	were designed, especially for high speed upset	pipes in Fig. 1.10). They further introduced transverse magnetizing concept system (Fig. 1.11) and its implication	
Edson [16]	• Reported the developments aimed at increasing the weldable wall thickness to about 12 mm (150KN-Machine Fig. 1.13) • Limitations that arise while welding tubes with higher wall thickness were highlighted	• The thickness limitation crops up from the arc rotating initially on the inside edge and then predominantly on the outer edge of the tube faces • As a result, uneven heating and consequent poor quality welds are obtained. This radial movement of the arc path occurs with any wall thicknesses but becomes less consistent as wall thickness increases above 5 mm • The second difficulty was to ensure immediate arc rotation in order to avoid local melting and resultant short circuiting	See Fig. 1.12

Fig. 1.12 150KN MIAB machine and monitoring equipment

1.3.3 Studies on Modelling and Simulation of MIAB Welding Process

Authors and year	Experimentation	Observations	Images and equations
Kim and Choi [3]	• Developed a two-dimensional finite element model for the analysis of magnetic flux density distributions produced by electromagnets at the MIAB weld joint • Their experimental set-up, shown in Fig. 1.13, utilized a Gauss-metre at the centre of the joint and the flux density was measured at a varying distance from the outer surface of the pipes at various distances from the exciting coil	• It is important to maintain maximum flux density at the joint for best weld quality. Therefore, the design of the electromagnet system is very important, as is the exciting current applied to the electromagnets • Both the gap size between the two pipes and the relative permeability of the medium therein had an effect on the magnetic flux at the joint	See Fig. 1.13
Vendan et al. [17]	• Performed non-linear electromagnetic analysis to determine the magnetic field and electromagnetic force distribution in MIAB process using finite element package ANSYS • Typical results of this analysis pertaining to magnetic field were compared with the experimental data for steel tubes (outer diameter 47 mm and thickness of 2 mm)	• The proposed three-dimensional finite element method model for electromagnetic force distribution facilitated comprehensive understanding of the arc rotation process in MIAB welding	NA

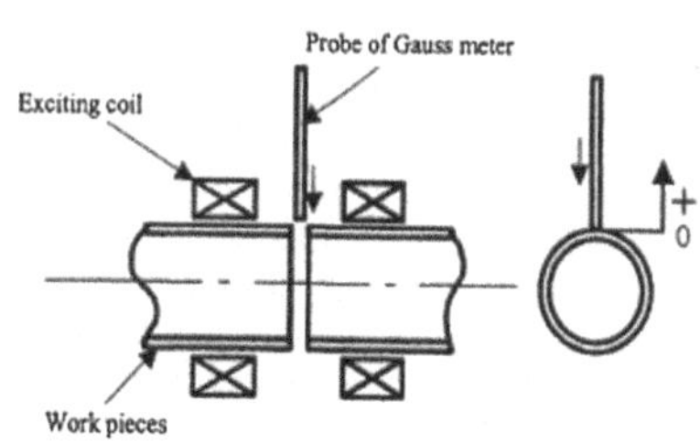

Fig. 1.13 Method for measuring magnetic flux density

1.3.4 Applications of MIAB Welding

Authors and year	Experimentation	Observations	Images and equations
Fletcher et al. [18]	• A prototype MIAB welding machine was designed and built which was capable of welding natural gas pipelines and to make welds in DN 150 pipe complying with the performance requirements of the Australian petroleum pipeline standard AS2885.2	• A typical MIAB welded pipe is shown in Fig. 1.14	See Fig. 1.14
Tagaki et al. [12]	• Developed an equipment for rotating arc butt welding suited to pipeline laying in urban areas. The machine was employed to weld pipes of 60.5 mm outside diameter with 3.8 mm wall thickness and 89.1 mm OD with 4.2 mm mild steel town gas pipelines	• It was shown that weld quality of high reliability can be obtained with high efficiency and that the welding equipment can be effective in pipeline laying	NA
Schlebeck [19]	• Presented the application of MBL-P (Magnetically Moved arc with pressure or MIAB) in engineering components such as CO_2 pressure cylinder for fire extinguishers, pipe screw joint for hydraulic lines and pipe/flange joints of 32–85 mm nominal width.	• Figure 1.15 shows the MBL welded CO_2 pressure cylinder for fire extinguishers for which the cycle time is 20 s. The test pressure amounts to 40 MN/mm^2 • Figure 1.16 shows a full-size MBL welded pipe screw coupling for a hydraulic line with operating pressures up to 15 MN/m^2	See Fig. 1.15 See Fig. 1.16
Westgate and Edson [20]	• Outlined the typical industrial applications within the automotive industry	• The application of MIAB welding to weld parts in Ford Transit car rear axle casing containing two circular and two square butt welds (Fig. 1.17)	See Fig. 1.17
Hagan et al. [21]	• Summarized their use of MIAB welding in the manufacturing of the Fiesta rear axle cross tube assembly • In selecting MIAB welding, they first considered the other more common welding methods, viz., Friction, Flash and GMAW	• Friction was not acceptable because of the difficulty in maintaining the radial relationship between the shaped flange spindles and the axle tube	

(continued)

(continued)

Authors and year	Experimentation	Observations	Images and equations
Hiller et al. [22]	• Used MIAB (in this case, Magnet arc) welding in the production of truck cab suspension components	• This application involved MIAB welding a cast iron lever to an extruded steel torsion tube to produce the welded assembly (Fig. 1.18) • The authors commented on the many advantages of this process, including short welding times and the excellent mechanical properties of the solid-state joint produced between cast iron and steel	See Fig. 1.18
Jenicek et al. [23]	• Demonstrated that tubular hollow bodies such as nuts, sleeves and bushes could be fastened to sheets using a process with particular economic viability, i.e. an advanced variant of magnetically impelled arc butt welding-bush or nut welding	• With extended drawn-arc stud welding devices, aluminium components with an internal thread between M8 and M24 were welded on to perforated sheets made of ENAW-$AlMg_3$ and ENAW-$AlMgSi_1$	NA
Mori et al. [24]	• Evaluated the feasibility of the MIAB welding process with aluminium and aluminium-copper joints • In this set-up, it was a challenge to achieve the required flux density at the joint with non-ferrous materials versus ferrous materials. Hence, an iron core was often inserted inside the pipe	• Results achieved showed good weld strength and metallurgical integrity	NA

Fig. 1.14 MIAB welded pipes

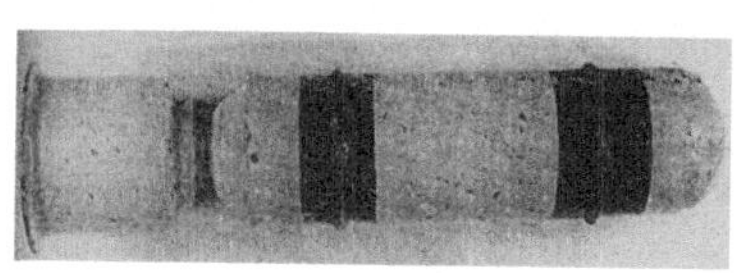

Fig. 1.15 MBL welded CO_2 pressure cylinder for fire extinguishers

Fig. 1.16 MBL welded pipe screw coupling for a hydraulic line

Fig. 1.17 Rear axle casing in Ford Transit car

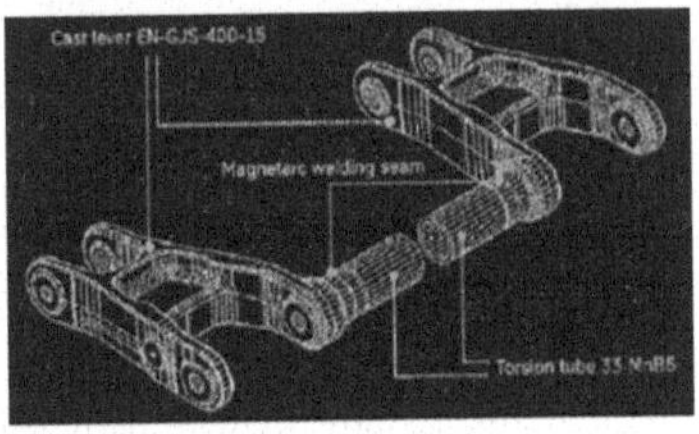

Fig. 1.18 MIAB (Magnet arc) weld joint of cast iron-to-steel [6]

1.4 Experimentation and Process Governing Parameters

Investigations are carried out on MIAB welding machine (MD1) after preliminary experimentation to understand the basic mechanisms involved in MIAB welding process. The experimental procedure involves a series of trials to develop and evaluate the knowledge base for MIAB welding alloy steel tubes. Based on the penetration and bead of the weld (observed through visual inspection), the appropriate ranges of various input process parameters are selected for conducting further experimental trials.

1.4.1 Miab Equipment

MIAB machine (MD1) available at welding research institute (WRI), BHEL, is capable of welding alloy steel tubes and is suitable for real-time application in industrial sectors.

MD1 MIAB machine consists of a power converter, pump station, control box, welder unit and the remote control as major blocks [Technical Manual (speed arc-1)] as shown in Fig. 1.19. The input to the power converter is 3ph, 50 Hz, 400 V.

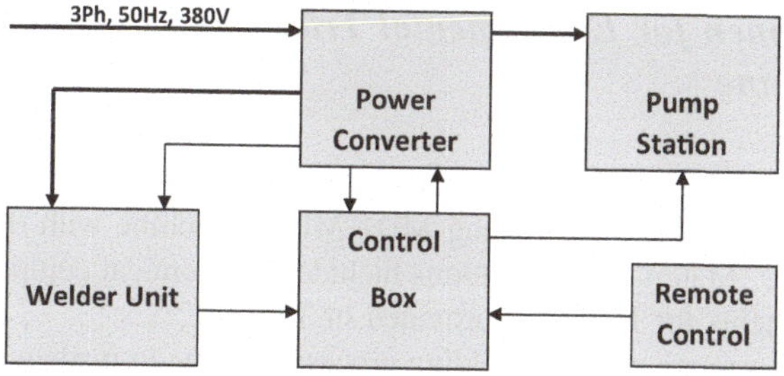

Fig. 1.19 Schematic diagram of MIAB welding machine MD1

The structure of the machine includes the following basic units:

- Welding unit
- Control cabinet
- Console
- Hydraulic pump station
- Die sets and
- The welding rectifier

The machine (Fig. 1.20) is intended for work indoors and also outdoors. The machine is complete with a set of cables connecting its base parts.

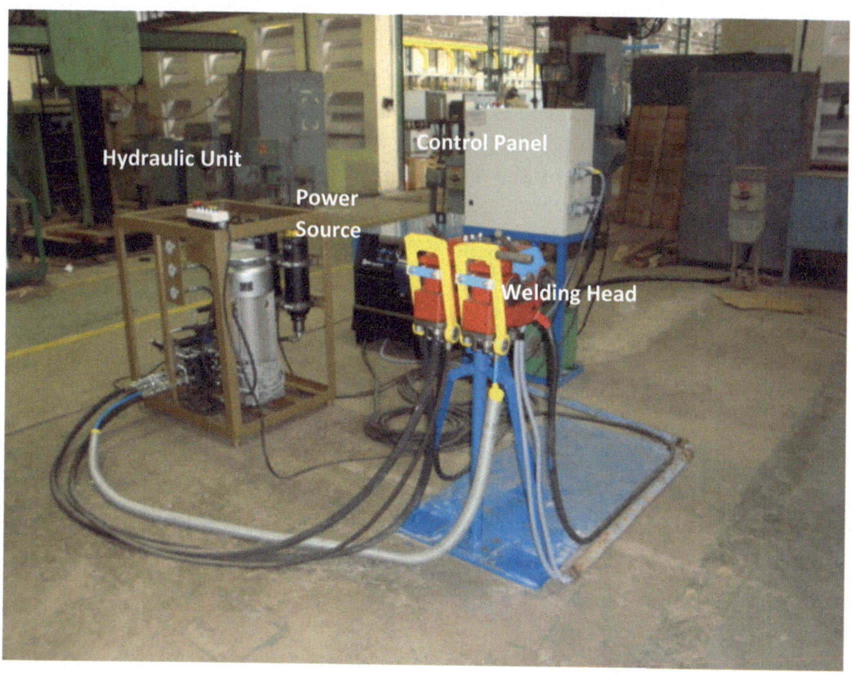

Fig. 1.20 MIAB machine (MD1) with the basic units

1.4.2 Specimen for Experimental Trials on MD1 MIAB Machine

Several welding trials are conducted with alloy steel tubes (T11) of 47.6 mm outer diameter and 6.6 mm thickness using MD1 MIAB machine with different sets of input parameters. Material specifications include the chemical composition and the tensile requirements for MIAB as provided in Table 1.2.

MIAB being a relatively new welding process and due to inadequate availability of literature, initially, trial and error method is adopted. The welding current and the welding time are divided into three and four stages, respectively. Tables 1.3 and 1.4 depict the connotation of the various stages of current and time along with the welding specifications recommended by the manufacturer of MD1 machine for welding a 47.6 mm diameter, 6.6 mm thickness steel tubes.

I_1 current is employed for around 0.5–1 s during which the tubes to be welded are shorted and then retracted when an arc is between the two tubes. For T11 composition, the range of current for the pre arc stage is 280–300A. The I_2 current is in the range 200–220 A during which the arc is impelled along the peripheral edges of the tubes while heating up the faying surfaces to the point of solidus (red hot condition). Finally, the I_3 current of about 1100 A during the upsetting of the two tubes whose faying surfaces are in red hot condition. A high current in stage III (also referred to as expulsion current) is expected to play an important role in the cleaning action by expelling the impurities when forging action is employed. These specifications of current are applicable only for welding T11 tubes of 47.6 mm diameter with 6.6 mm thickness.

MIAB welding process is divided into four stages based on the time as specified by the manufacturer. T_1 (0.5–1 s) is the time during which the tubes, initially with a gap between them are shorted. T_2 (4–6 s) is the time period in which the shorted tubes are retracted followed by the arc striking between the tubes. Further, in this

Table 1.2 T11 Material specifications [25]

Mat.	C	Mn	P_{max}	S_{max}	Si	Cr	Mo	R_m (MPa) T.S ksi	R_p 0.2 (Mpa) Y. S ksi	A (%) Elong
SA213–11	<0.15	0.3–0.6	0.03	0.03	0.5–1.0	1.00–1.5	0.44–0.65	415–460	205–230	30

Table 1.3 Range of current for various stages

Stages Nos.	Operation	Current range in (A)
I_1	Pressing of tubes and arc initiation	280–300
I_2	Arc heating	200–220
I_3	Upset	1100

Table 1.4 Range of time period for various stages

Stages nos.	Operation	Time range (s)
T_1	Pressing of tubes	0.5–1
T_2	Tube separation, arc initiation and start of arc rotation	4–6 s
T_3	Arc heating of butt ends-stabilized rotation of arc	20–30
T_4	Upset	0.3

stage of time, the arc starts to slowly rotate along the peripheral edges of the tube ends. During this stage of time, I_1 current of (280–300 A) is employed. In stage T_3 (20–30 s), arc accelerates and rotates with a high speed of about 250 m/s (linear) along the peripheral edges of the tubes while heating up the faying surfaces to the point of solidus (red hot condition). A current of 200–220 A (I_2 current) and is provided during the T_3 stage. Finally, upsetting action takes place in T_4 (0.3 s) by supplying a current of 1100 A (as per I_3 stage).

1.4.3 Trials Conducted Using MD1 Machine

Initially, trial and error method is adopted to evaluate the impact of various input parameters on the penetration and bead of the weld. From the specifications provided by the manufacturers for welding 47.6 mm diameter 6.6 mm thickness, it is noticed that the I_2 of current and T_3 of time play vital role in deciding the penetration and bead of the weld as these stages are employed for a longer duration. However, to determine the nature of impact of various input parameters on the penetration and weld bead, slight variations are made in the values of the process parameters deviating from the manufacturers' specifications.

Trials are carried out varying the settings of current and time during different stages while assuring that the deviation from the manufacturer's specification is minimal. The I_2 current is the only parameter that is varied to the maximum extent as it plays a significant role in MIAB welding process. These trials are carried out mainly to identify appropriate levels of various input parameters that lead to uniform bead formation with sufficient penetration. For trials 1–19, the I_2 current is maintained at 100 A. The I_1 current is varied in the range 280–300 A for various trials. The distance between the two tubes (arc gap) is varied between 4 and 2 mm for different trials. This is to arrive at the suitable value of the distance needed to achieve weld without spending excessive energy. The I_3 current is maintained constant for most of the trials at 1100 A as specified by the manufacturer. The T_1 and T_4 stages of time are maintained constant at 1 and 0.3 s for all the trials. The T_2 and T_3 stages of time are varied from 4–6 s to 20–30 s for the various trials. The upset pressure is kept constant at 150 N/mm^2 for all the trials. The values of the parameters are randomly varied during the trials and the observations made through visual inspection during various trials are documented in Table 1.5. Pictures of welded tubes (typical) with bead profiles are shown in Fig. 1.21.

Table 1.5 Experimental trials conducted on 47.6 mm diameter, 6.6 mm thickness T11 tubes

Trial no.	Current (A)			Time in (s)				Upset pressure (N/mm^2)	Distance between tubes (mm)	Remarks (visual inspection)
	I_1	I_2	I_3	T_1	T_2	T_3	T_4			
1	280	100	1100	1	4	20	0.3	150	3	No weld
2	280	100	1100	1	4	25	0.3	150	3	No weld
3	280	100	1100	1	4	30	0.3	150	3	No weld
4	280	100	1100	1	6	20	0.3	150	3	No weld
5	280	100	1100	1	6	25	0.3	150	3	No weld
6	280	100	1100	1	6	30	0.3	150	3	No weld
7	300	100	1100	1	4	20	0.3	150	3	No weld
8	300	100	1100	1	4	25	0.3	150	3	No weld
9	300	100	1100	1	4	30	0.3	150	3	No weld
10	300	100	1100	1	6	20	0.3	150	3	No weld
11	300	100	1100	1	6	25	0.3	150	3	No weld
12	300	100	1100	1	6	30	0.3	150	3	No weld
13	280	100	1100	1	6	20	0.3	150	1	Tubes getting shorted
14	280	100	1100	1	4	20	0.3	150	1	Tubes getting shorted
15	280	100	1100	1	4	20	0.3	150	2	Incomplete penetration
16	280	100	1100	1	4	25	0.3	150	2	Incomplete penetration
17	280	100	1100	1	4	30	0.3	150	2	Incomplete penetration
18	280	100	1100	1	6	20	0.3	150	2	Incomplete penetration
19	280	100	1100	1	6	30	0.3	150	2	Incomplete penetration
20	280	200	1100	1	4	20	0.3	150	2	Good weld
21	280	200	1100	1	4	22	0.3	150	2	Good weld
22	280	200	950	1	4	22	0.3	150	2	Achieved weld, non-uniform bead, very less flash and not full penetration
23	280	200	1000	1	4	22	0.3	150	2	Achieved weld, non-uniform bead, very less flash and not full penetration
24	280	300	1100	1	4	22	0.3	150	2	Achieved weld, excess material expulsion
25	280	300	1100	1	6	20	0.3	150	2	Achieved weld, excess material expulsion
26	280	300	1100	1	4	18	0.3	150	2	Good weld. Uniform penetration

a) MIAB welded tubes

b) Tulip test sample for weld integrity assessment

c) MIAB welded joint depicting a uniform external bead profile

d) MIAB welded joint depicting a uniform internal bead profile

Fig. 1.21 MIAB welded joint characteristics in terms of bead profile

The results from Table 1.5 show that MIAB welding of T11 tubes (47.6 mm dia, 6.6 mm thickness) with a I_1 current of 100 A is not possible. Though, manufacturer clearly specifies a I_2 current of 200 A for welding T11 tubes, a comparatively lower I_2 current is supplied in order to check the possibility of welding T11 tubes for that current which would reduce the energy consumption. The other parameters such as I_1 current, distance between the tubes and various stages of time are varied when 100 A (I_2 current) is supplied. However, still it is evident that at 100 A current T11 tubes cannot be welded. The trials from 1 to 19 showed that it is always advisable to maintain a distance of about 2 mm between the tubes for accomplishing the weldment. When the distance is maintained at 4 mm and 3 mm, respectively, there is no weld. On the contrary, when 1 mm distance is maintained between the tubes, the tubes are getting shorted and arc is getting quenched. The trials (1–19) provide guideline regarding the distance that needs to be maintained between the tubes to attain a weld.

Fig. 1.22 Uniform bead and full penetration

Subsequently, I_2 current is raised to 200 A for the trials (20–23). It is observed that when the (T_2) is 20 and 22 s a good weld is obtained when visually inspection is performed as shown in Fig. 1.22, while maintaining other parameters in the range specified by the manufacturer. However, when the I_3 current is reduced in trials (22 and 23) to 950 and 1000, weld with non-uniform bead, very less flash and insufficient penetration is observed. This emphasizes the significance of I_3 current which governs the last stage of the MIAB welding process, though it comes into effect only for a shorter duration. Hence, it is required that the I_3 current must be maintained at 1100 A.

After the trials (1–23), further trials are conducted by increasing the I_3 current to 300 A. Excess material expulsion and non-uniform bead are observed in the weld obtained for trials (24 and 25). This may possibly be due to the reason that a current of 300 A for an arc heating time of 22 and 20 s causes excessive heating of tube edges leading to melting of tubes and expulsion of material during forging as shown in Fig. 1.23. Thus the main characteristic feature of solid-state welding of MIAB

Fig. 1.23 Excess melting and flashing

Table 1.6 Additional tests on 47.6 mm diameter, 6.6 mm thickness T11 tubes based on (2 × 5 design matrix)

Sl. no	Material and identification mark	Current (A)			Time (s)				Upset pressure (N/mm^2)	Distance between tubes (mm)
		I_1	I_2	I_3	T_1	T_2	T_3	T_4		
1	T11/1	280	200	1100	1	6	14	0.3	150	2
2	T11/2	280	200	1100	1	6	17	0.3	150	2
3	T11/3	280	200	1100	1	6	20	0.3	150	2
4	T11/4	280	200	1100	1	6	23	0.3	150	2
5	T11/5	280	200	1100	1	6	26	0.3	150	2
6	T11/6	280	250	1100	1	6	14	0.3	150	2
7	T11/7	280	250	1100	1	6	17	0.3	150	2
8	T11/8	280	250	1100	1	6	20	0.3	150	2
9	T11/9	280	250	1100	1	6	23	0.3	150	2
10	T11/10	280	250	1100	1	6	26	0.3	150	2

welding process is absent. As in solid-state welding processes, the material is not taken to the melting temperature. Consequently, the arc heating time is reduced to 18 s in the 26th trial. As expected, there is no melting of material and a good weld is obtained.

Trials (1–26) are carried out varying the values of the parameters over a narrow range around the manufacturer's recommendations to get a few fundamental ideas about the process parameters governing MIAB welding process. With the primary and essential knowledge from the trials conducted based on trial and error method, further trials are conducted. However, subsequent trials are conducted according to a 2 × 5 design matrix. Two levels of current I_2 and five levels of arc heating time T_3 are considered for these trials. The other parameters are employed only for a short duration and hence are less significant and are maintained at levels specified by the manufacturer (similar to that of the trials 20 and 21). The design matrix with two levels of I_2 current and five levels of T_3 arc heating time are shown in Table 1.6. The specimens welded as per the parameter settings provided in Table 1.6 for each trial are then tested for mechanical strength and metallurgical properties of the weld. The surface of tubes before welding is cleaned to avoid impurities at the joint.

1.4.4 ARC Monitoring During MIAB Welding

During the MIAB welding process, the arc is monitored using arc monitoring equipment. The data of current and voltage are analysed using the software adaq 3000. The data acquisition system facilitates recording data of voltage and current of the arc for 4 s. The arc monitoring method and the recorded current and voltage are shown in Figs. 1.24 and 1.25, respectively. This recording is for the 20th trial in Table 1.6 which wielded a good weld in terms of penetration and bead as visually observed.

Fig. 1.24 Arc monitoring during MIAB welding

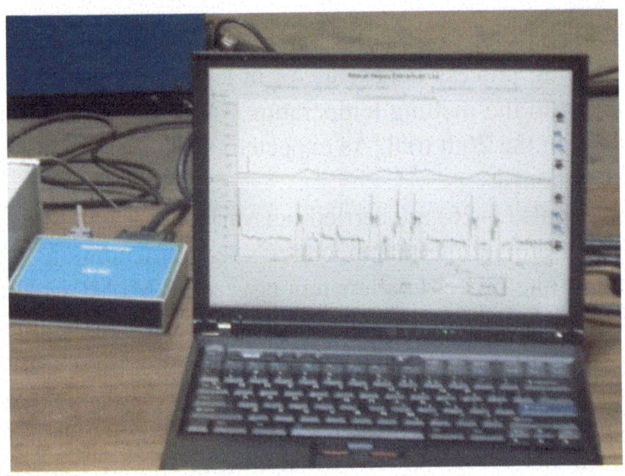

Fig. 1.25 Recorded voltage and current of the arc

Arc voltage and current during the time interval 0–4 s is shown in Fig. 1.26. Initially, there is a steep drop in the voltage (almost zero) as the two tubes are pressed together and the current is high (about 350 A).

Further, when the two tubes are retracted to a distance of 2 mm an arc is initiated. This voltage of the arc fluctuates while establishing and rises to about 60 V before stabilizing. The voltage of arc is around 25 V after stabilization. The current of the arc remains constant at about 200 A. The current and voltage of the arc during the time interval 14–18 s are illustrated in Fig. 1.27. The current and voltage of the arc during this stage remain constant at around 200 A and 25 V, respectively. In this time period, the arc revolves with a constant speed and uniform heat is dissipated from the arc to the tube edges.

Figure 1.28 demonstrates the voltage and current characteristics of the arc during the final stage of the MIAB welding process (22–26 s). It is observed that for about 40 ms of the 4 s period of recording, the arc draws a current of about 1050 A and

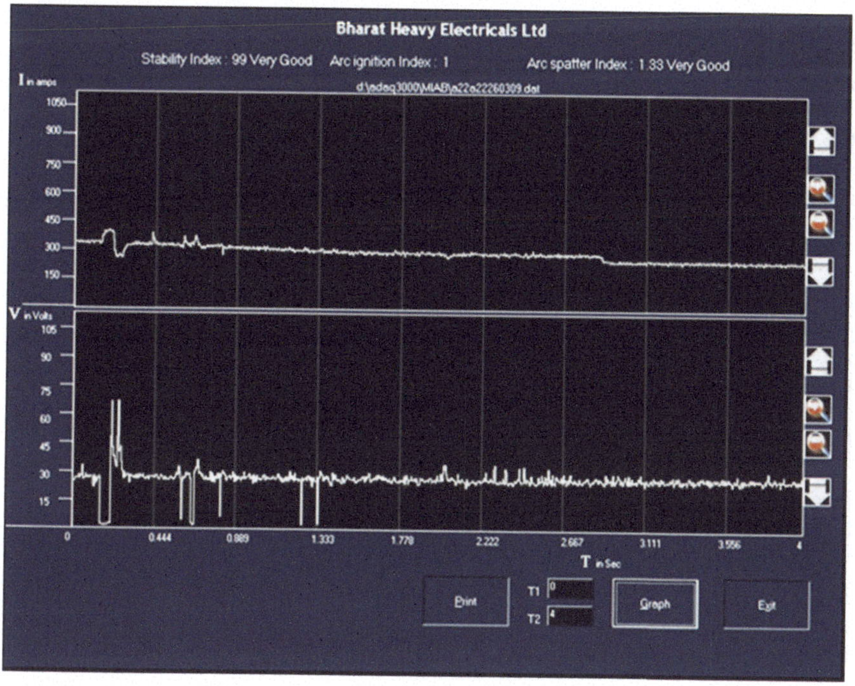

Fig. 1.26 Voltage and current characteristics of arc during 0–4 s of MIAB welding

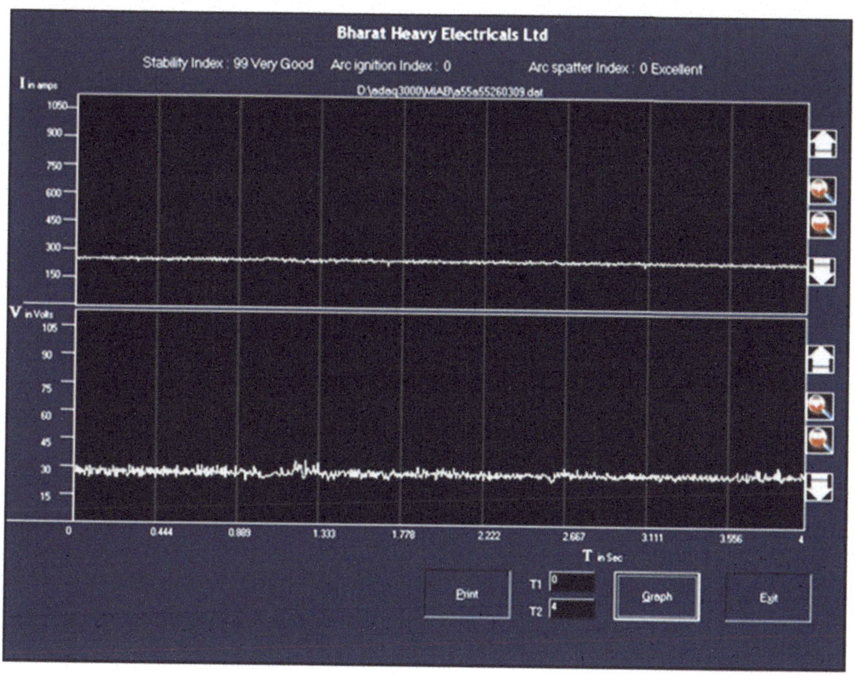

Fig. 1.27 Voltage and current characteristics of arc during 14–18 s of MIAB welding

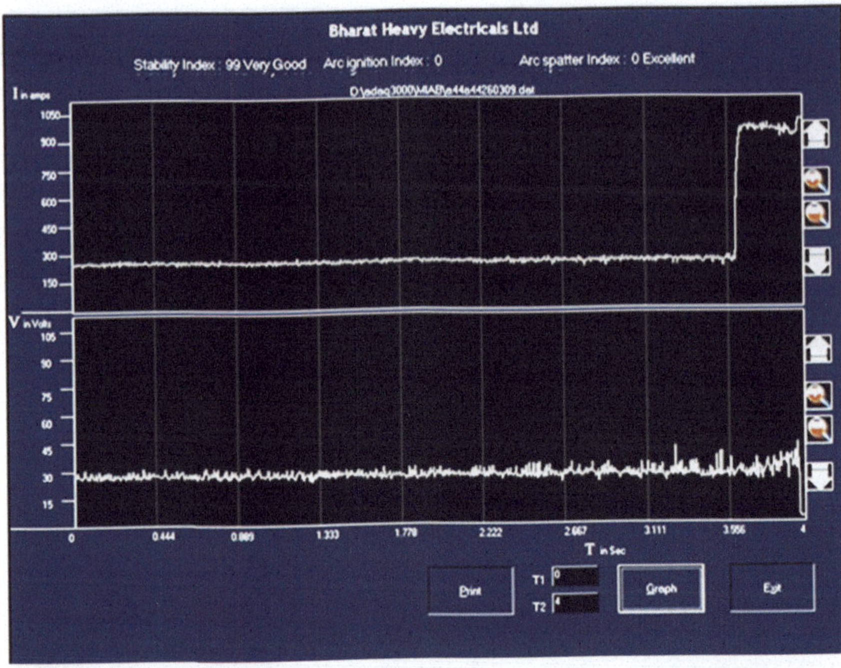

Fig. 1.28 Voltage and current characteristics of arc during 22–26 s of MIAB welding

there is notable fluctuation in the voltage just before the forging. Finally, the voltage drops to zero when the tubes are forged together.

1.4.5 Inferences

Experimental trials are carried out initially using trial and error basis on T11 tubes (47.6 mm diameter and 6.6 mm thickness). The process parameters are varied while checking the impact on the weld bead and penetration (examined visually). The uniformity in the bead and the expulsion of the flash are the quality parameters for the visual inspection. The following conclusions are made:

- The I_2 current employed for a longer duration plays a significant role in governing the weld bead and penetration. The I_2 current of 100 A is found to be inadequate as it does not produce necessary heating at the faying surfaces. This leads to insufficient penetration and results in either poor weld or no weld. I_2 current of 200 A is found to yield a good quality weld for T_3 time of 20 and 22 s. Similarly, I_2 current of 300 A also results in a full penetration and uniform bead for a T_3 time of 18 s. However, I_2 current must be minimized in order to reduce the input energy consumed during the welding process. Hence, a current of 200 A is most appropriate for achieving the welding of T11 tubes (47.6 mm dia, 6.6 mm thick).

- It is imperative to maintain a high current of 1100 A during the last few milliseconds just before forging to get a good quality weld.
- The gap between the tubes is to be maintained at 2 mm to obtain a good weldment. A gap of below 2 mm between the tubes leads to shorting of the tubes when arc is initiated or quenched.

With the basic knowledge on the ranges of input parameters, a 2 × 5 matrix is designed with two levels of I_2 current and four levels of time. Trials are carried out based on the design matrix and the welded specimens are tested for mechanical and metallurgical properties. Further, the characterization of materials is expected to reveal the appropriateness of the ranges for the input parameters for the MIAB welding process.

1.4.6 Results and Discussions

1.4.6.1 Evaluation of Mechanical Strength Through Destructive Testing

MIAB welding trials are carried out on T11 tubes for different sets of process parameters. Trials are carried out as per the design matrix shown in Table 1.6. The surface of tubes is cleaned before welding to avoid impurities at the joint. Initially, a visual inspection is carried out on the MIAB welded tubes for the ten sets of process parameters. For these trials, the distance between the tubes is maintained at 2 mm. The general observations for each trial are provided in Table 1.7. It is seen that a uniform bead with full penetration is achieved in case of the trials 3, 4, 7 and 8. This gives the idea of the range of arc heating time that must be maintained for a full penetration. The arc heating time of about 17–23 s seems to yield a full penetration with a uniform bead along the weld joint for a stage II current of 200 A and 300 A, respectively. However, trial 1 yields a weld with insufficient penetration but uniform reinforcement and trial 2 yields a full penetration with non-uniform reinforcement, respectively, for an I_2 current of 200 A. Contrary to this, trials 9 and 10 yield full penetration. However, excess loss of material takes place due to melting of the butt ends. Consequently, the reinforcement height increases to about 3.5–4.5 mm on the outer surface and 2–2.5 mm on the inner surface for the 9th and 10th trial, respectively, for 300 A I_2 current. Further, for the 10th trial, when excess metal is collected underneath the weld (due to gravity), it leads to certain undesired situations such as short circuiting the tube gap region, quenching the arc or a local explosion. Any of these situations can thus result in an unfilled gap on the weld line on forging. Such weld specimens are likely to fail in the ball test. Thus, only the specimens from trials 1, 2, 3, 4, 7 and 8 are considered for various destructive tests to determine their strength.

Table 1.7 Trials conducted with 2 levels of I_2 current and 5 levels of T_3 time

Trial. no	Specimen/ identification mark	Current (A)			Time (s)				Observations (visual inspection)
		I_1	I_2	I_3	T_1	T_2	T_3	T_4	
1	T11/1	280	200	1100	1	6	14	0.3	Insufficient penetration, uniform reinforcement
2	T11/2	280	200	1100	1	6	17	0.3	Full penetration Non uniform reinforcement
3	T11/3	280	200	1100	1	6	20	0.3	Full penetration, uniform weld bead
4	T11/4	280	200	1100	1	6	23	0.3	Full penetration, uniform weld bead
5	T11/5	280	200	1100	1	6	26	0.3	Melting of material, excess metal expulsion at the joint
6	T11/6	280	250	1100	1	6	14	0.3	Incomplete penetration
7	T11/7	280	250	1100	1	6	17	0.3	Full penetration, uniform weld bead
8	T11/8	280	250	1100	1	6	20	0.3	Full penetration, uniform weld bead
9	T11/9	280	250	1100	1	6	23	0.3	Melting of material, excess metal expulsion at the joint
10	T11/10	280	250	1100	1	6	26	0.3	Excess melting of material, gap at weld

Results of Tensile Test

The transverse tensile tests are conducted on five MIAB welded specimens and the fractured specimens are shown in Fig. 1.29. The tensile test results are given in Table 1.8 which indicates that the weld strength is slightly above the ultimate tensile strength in the base material of the specimen. Failure in the base material of the specimens in trials 3, 4, 7 and 8 indicate that the weld joint is stronger than the base material. However, failure of specimen in the weld metal in trial 2 indicates that the weld is weaker than the base material. This may be possibly due to the insufficient arc heating time.

Fractography

Figure 1.30 shows the MIAB welded specimen failing the tensile test with a fracture at the weld joint.

Scanning electron microscope (SEM) is used to study the fractured surface of weld specimen which failed the tensile test. The SEM photographs at magnifications 5000X are shown in Fig. 1.31, respectively.

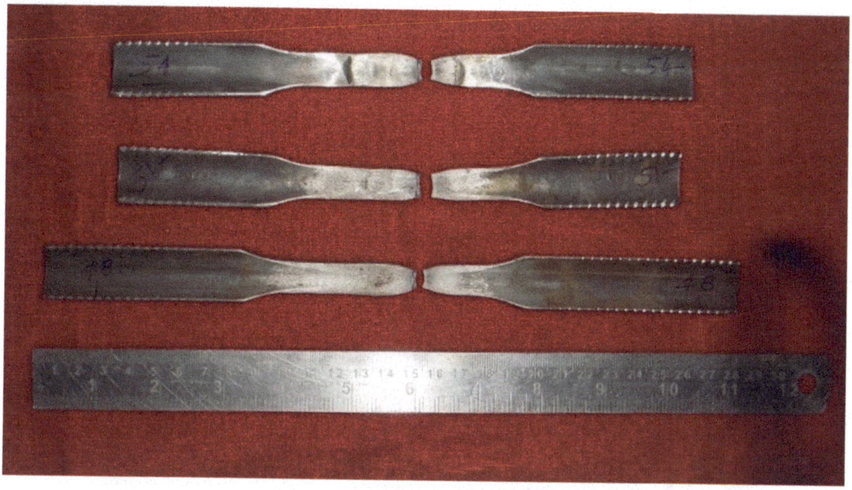

Fig. 1.29 Typical tensile tested MIAB weld tube samples

Table 1.8 Results of tensile test on the MIAB weld specimen

Specimen/material	Specimen size in mm	U.T.S in MPa	Position of fracture
T11/2	12.72 × 6.15	456	Weld metal
T11/3	13.05 × 6.00	506	Base metal
T11/4	12.59 × 6.00	498	Base metal
T11/7	12.43 × 6.00	493	Base metal
T11/8	12.51 × 6.10	448	Base metal

Fig. 1.30 Tensile test Specimen—fracture in weld metal

From Fig. 1.31, it is observed that a brittle fracture occurs at the failing tensile test at weld region. This brittle fracture is defined as a fracture which takes place due to the rapid propagation of crack with a negligible deformation. This may possibly be due to inadequate heat input or heat for a short input time. The solid-state deformation in the tube edges may be sufficient. Further, the strain rate may have increased in the joint. The crack length may be too small leading to the development of higher stress at the weldment and thereby to the brittle fracture.

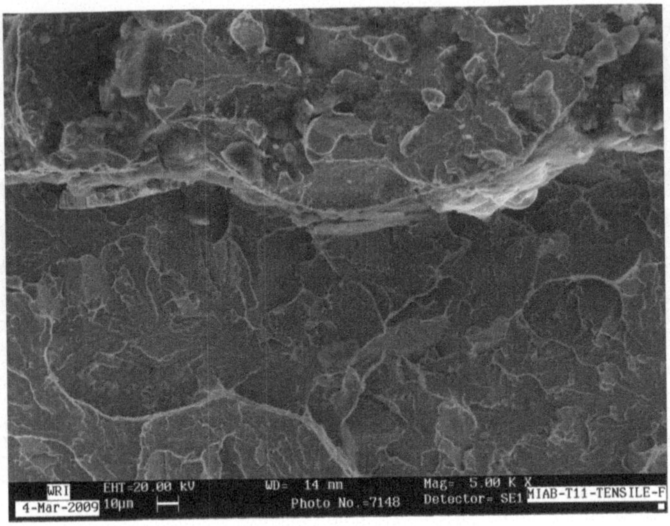

Fig. 1.31 SEM fractography of tensile failure specimen at 5000X

Results of Bend Test

For conducting the root bend test and face bend test, five specimens (each test) are considered. These specimens are labelled as 2R, 3R, 4R, 7R, 8R and 2F, 3F, 4F, 7F, 8F, respectively, for root bend test and for face bend test. The results of bend test are given in Table 1.9.

Figure 1.32 shows a few specimens subjected to the root bend test, while Fig. 1.33 shows a few specimens subjected to the face bend test. In both the tests, the specimens have no open discontinuity. It is observed that the specimens 2, 3 and 4 passed the bend tests and showed no or 1 mm open discontinuity both in the case of root and face bend tests, respectively. However, the specimens 7 and 8 fail in the

Table 1.9 Bend test results

Idfn.	Root bend	Remarks	Idfn.	Face bend	Remarks
T11/2R	No open discontinuity observed	Passed	T11/2F	No open discontinuity observed	Passed
T11/3R	No open discontinuity observed	Passed	T11/3F	No open discontinuity observed	Passed
T11/4R	1 mm open discontinuity observed	Passed	T11/4F	No open discontinuity observed	Passed
T11/7R	9 mm open discontinuity observed	Failed	T11/7F	No open discontinuity observed	Passed
T11/8R	15 mm open discontinuity observed	Failed	T11/8F	No open discontinuity observed	Passed

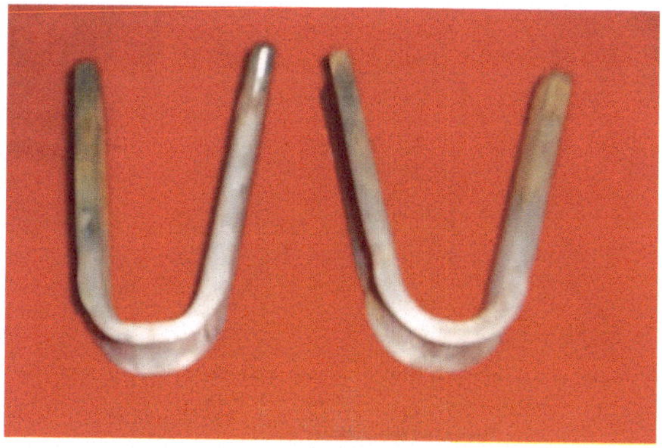

Fig. 1.32 Test specimens after root bend test

Fig. 1.33 Test specimens after face bend test

root bend test with an open discontinuity of about 9 mm and 15 mm, respectively. The general causes leading to a failure at the root bend may be insufficient forging pressure, varying tube gap distance or presence of impurities due to improper cleaning. In this case, the failure of root bend may be due to excess melting of metal and increase of brittleness and hardness at the weld. The angle of bend used for this study is 180° while the mandrel diameter is 4t (where "t" thickness of the specimen).

Microhardness Survey

In this test, the hardness is measured using the Vickers microhardness tester by computing the depth of penetration of a penetrator into the specimen under certain fixed condition of test. The load used is 0.5 kg and the magnification is 400×. A distance of 0.8 mm is considered for each reading while measuring the microhardness. Figures 1.26, 1.27, 1.28, 1.29 and 1.30 show the microhardness in MIAB welded specimens 2, 3, 4, 7 and 8 in horizontal and vertical directions, respectively, for various I_2 current and T_3 arc heating time. Figure 1.34a illustrates that the

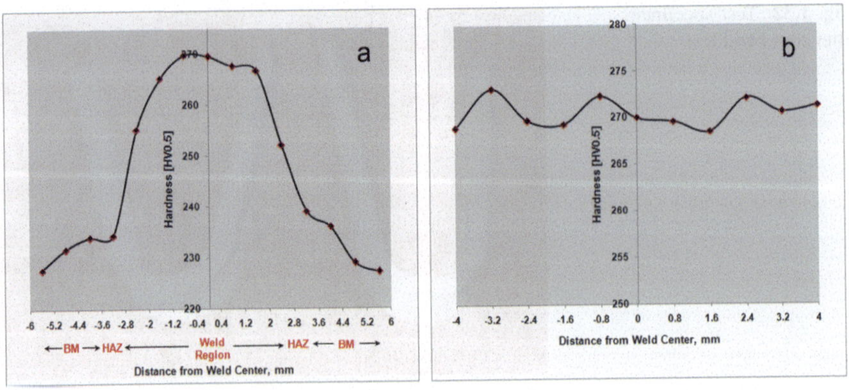

Fig. 1.34 **a** Microhardness across the weld line for specimen 2, **b** Microhardness along the weld line for specimen 2

microhardness levels are high across the weld line for all the cases. It varies between 265 and 300 VHN. The hardness in the heat-affected zone (HAZ) is around 230–260 VHN. The hardness in the base metal is around 210–235 VHN. From the microhardness test, the hardness in base metal, HAZ and the weld region are determined. The weld has good strength because the hardness in the weld zone is higher than the hardness in base metal.

Results of Impact Test

Impact testing involves subjecting the specimens (weld metal at the centre) and base metal to impulsive or suddenly applied loads. The specimens are tested at 0° and the results obtained are presented in Table 1.10. The impact energy values for base metal are the average of three specimens.

The test results indicate variation of impact strength for base material, welds with varying heat input. The impact toughness at 0° is found to be nearly 26 J or the base metal. The impact strength for specimens 2, 3, 4, 7 and 8 are 27, 28, 24, 23 and 21, respectively. It is notable that the impact toughness decreases with an increase in heat input as in specimens 3, 4, 7 and 8. Further, it is observed that specimen 2

Table 1.10 Impact test results

Specimen	Energy absorbed (J)
Base metal	26
2	27
3	28
4	24
7	23
8	21

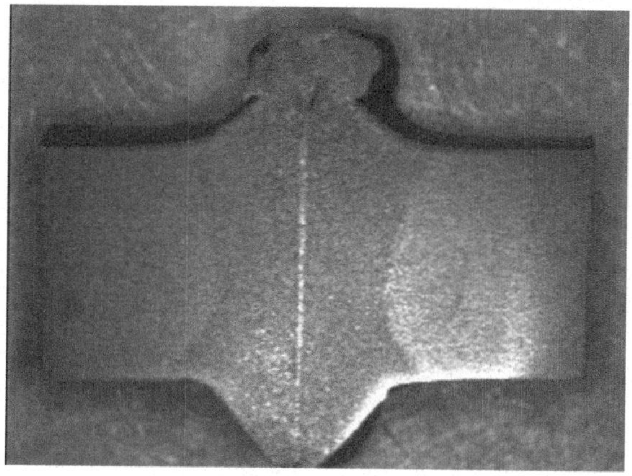

Fig. 1.35 Macrograph of MIAB weld (Specimen 2)

yields the highest impact strength. Since test standards are not yet available for MIAB welded specimens, the pass or failure of specimen for impact test cannot be specified.

Macroanalysis

The macrographs for the specimens 2 and 3 are presented in Fig. 1.35.

From the macrographs, the weld is distinct with solidification lines clearly visible. The centre line mark indicates the fused mass of materials. The inner diameter side is having more plastic bonded zone compared to outer diameter side.

Microanalysis

From the micrographs (Fig. 1.36), it is observed that no distinct HAZ or fusion line is visible. It is merged with the weld metal due to higher rate of heating (instantaneous higher heat input) and subsequent higher rate of cooling (due to very low weld cycle, i.e. about 25 s for the sample) and narrow HAZ. During heating, the overheating area is formed in the spots of the burning welding arc. During the upset, the coarse-grain area is pressed out into pre-external reinforcement and flash and the welded joint is formed at the expense of the fine-grain area.

Results of Radiography Test

The SA213A T11 tubes are grinded, then buffing is performed and the butt surfaces are cleaned to remove the impurities. Consequently, the tubes are MIAB welded (parameters as per trial 20 of Table 1.5 and radiography test is performed on the weld joint after removal of the reinforcement. The result is shown in Fig. 1.37. It is

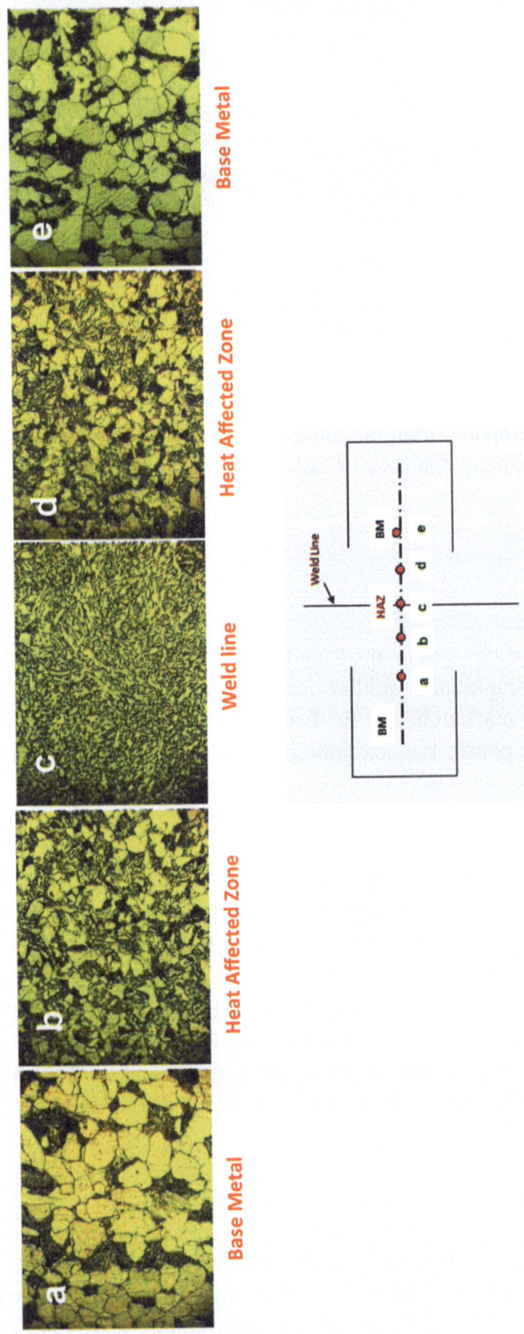

Fig. 1.36 Microstructures for specimen 2

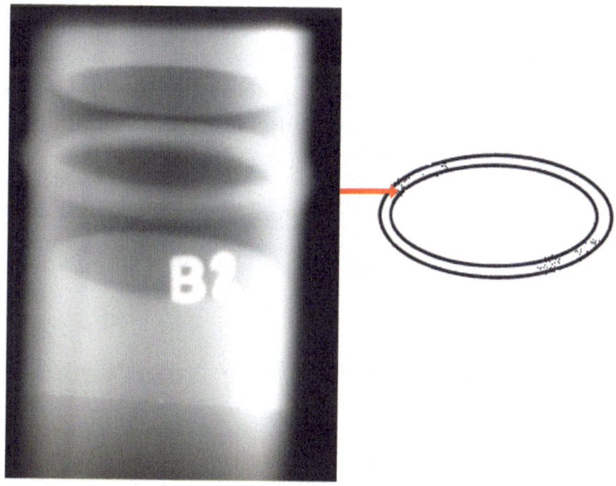

Fig. 1.37 X-ray film indicating the presence of pores at the joint (without reinforcement)

observed that there are negligible pores. This concentration of pores is within the acceptable limits as per ASME section 5. Thus the study demonstrates that the butt surface of the tubes must be free from impurities to get a better result for radiography test. These small number of pores may arise due to the interaction of the gas molecules in the weld region during the heating and cooling process of welding.

1.5 Parametric Analysis Using Machine Learning Terminologies

Various inference algorithms, which simplifies a mathematical function into a familiar form are called parametric machine learning algorithms. Some of the popular parametric machine learning algorithms extensively used in various applications are

1. Linear Regression
2. Logistic Regression
3. Linear Discriminant Analysis
4. Perceptron
5. Naive Bayes
6. Simple Neural Networks

The factors that lead to the consideration of parametric machine learning algorithms to infer about the efficiency of various welding methodologies are as follows:

1. **Straightforward**: Easy to understand and interpret both the algorithm and results
2. **Speed**: Inference rate of parametric models are fast enough
3. **Less Data**: These algorithms require less amount of data to train their model [6].

The sections mentioned below mainly focus upon achieving effective data analysis with the help of Python programming language along with its specific tools and libraries ecosystem.

1.5.1 Data Analysis Using NumPy *and* Pandas

This section gives a brief overview about some popular numerical computing and analysing Python packages.

NumPy an acronym for Numerical Python is an essential package used for efficient numerical computation in Python. The package constitutes of algorithms, data structures and library which plays a vital role in developing scientific applications involving extensive usage of mathematical data in Python.

The package also contains some other essential tools that aids in performing computation task for the applications are as follows:

- Quick and competent multidimensional array object ndarray, mostly used for faster mathematical operations.
- Efficient mathematical function to achieve enhanced and fast operations on the whole array set of data.
- Robust tools for performing reading and writing operation on datasets which are array based on the storage disk of a computing system.
- Provides support for complex mathematical operations related to random number generation, Fourier transformations and Linear algebra.
- A completely evolved C API for establishing errorless association between NumPy and libraries developed in legacy platforms such as C, C++ or FORTRAN [26].

The following section discusses the elements in detail.

Pandas is one of the most important tools widely used for faster data preprocessing and analysis. It is embedded with advanced data structures and data manipulation tool specifically designed for programmers to perform faster data analysis and cleaning operations in Python. The data structures used by Pandas for faster analysis are *Series* and *Data Frame*.

Through the entire book, frequent discussions on importing Pandas are encountered and the occurrence is in the following manner as shown in Fig. 1.38.

Fig. 1.38 Instruction to import Pandas

```
import pandas as pd
```

Fig. 1.39 Instruction to import data structures

```
from pandas import Series, DataFrame
```

Since the data structures are going to be frequently used, importing them into the local namespace is vital and can be achieved in the following ways (Fig. 1.39).

Understanding Pandas requires fundamental prerequisites about its widely used data structures such as Series and Data Frames. Individually discussions about each of these data structures will follow.

The datasets MIAB_NTS, MAIB_UTS, MAIB_HD are used to predict Notch Strength Ratio, Ultimate Tensile Strength and Mean Weld Interface Hardness. The first step is to import the dataset. The figures shown below give an idea on how to import the dataset using Pandas (Figs. 1.40, 1.41 and 1.42).

1.5.2 Data Visualization Using Seaborn *and* Matplotlib

"Data is the new oil" has become a new slogan in this era of data analytics and artificial intelligence. Producing informative results of the data in the form of plots is one of the significant tasks in data analysis. Efficient Python libraries like seaborn and matplotlib to generate static or dynamic visualization have been adopted.

The basic prerequisite for visualization is data. The data can be obtained in the form of built-in datasets or Pandas Data Frame. It is initiated by importing a built-in datasets along with libraries that would aid to seaborn for visualization (Fig. 1.43).

The analysis and visualization done for the datasets are to find out the co-relation between the independent variables, i.e. inputs and dependent variable, i.e. output by the help of seaborn library. Figure 1.44 shows the heat map and co-relation matrix for each dataset.

From the co-relation matrix of MIAB_NTS mentioned above, the following input parameter such as I_1 and I_3 are positively co-related to NTS, i.e. Notch Strength Ratio (Fig 1.44).

From the co-relation matrix of MIAB_UTS mentioned above, the following input parameter such as I_2 and I_3 are negatively and positively co-related to UTS, i.e. Ultimate Tensile Strength, respectively (Fig. 1.45).

From the co-relation matrix of MIAB_HD mentioned above, the following input parameters such as I_3 is positively co-related to Hardness (Fig. 1.46), i.e. Ultimate Tensile Strength.

1.5.3 Data Preprocessing Using Scikit-Learn

Data pre-processing is one of the crucial steps in machine learning. This is an unavoidable step in case of data which are in unstructured form. Further to this, we

```
# Load the data and replace the '..' with nan
ts_df = pd.read_excel('MIAB_NTS.xlsx')

ts_df = ts_df.astype(float)

display(ts_df)
```

	I1	I2	I3	t2	t3	NTS
0	330.0	310.0	1000.0	6.0	12.0	726.0
1	300.0	280.0	600.0	6.0	8.0	562.0
2	330.0	280.0	1000.0	4.0	12.0	705.0
3	330.0	310.0	1000.0	4.0	8.0	642.0
4	330.0	280.0	600.0	6.0	12.0	619.0
5	300.0	310.0	600.0	4.0	8.0	570.0
6	300.0	310.0	1000.0	6.0	8.0	713.0
7	330.0	280.0	1000.0	6.0	8.0	717.0
8	300.0	310.0	600.0	6.0	12.0	553.0
9	300.0	310.0	1000.0	4.0	12.0	708.0
10	300.0	280.0	1000.0	6.0	12.0	726.0
11	330.0	310.0	600.0	6.0	8.0	642.0
12	330.0	280.0	600.0	4.0	8.0	591.0
13	300.0	280.0	600.0	4.0	12.0	634.0
14	300.0	280.0	1000.0	4.0	8.0	675.0
15	330.0	310.0	600.0	4.0	12.0	557.0

Fig. 1.40 Importing MIAB_NTS

will be discussing about data preprocessing by applying machine learning libraries such as scikit-learn [6].

Polynomial features

This is a preprocessing technique used to derive non-linear features. It is mostly applied along with linear regression algorithm to train the model of higher degree. It is implemented in the following manner (Figs. 1.47, 1.48 and 1.49).

```
# load the data and replace the '..' with nan
ts_df = pd.read_excel('MIAB_UTS.xlsx')

ts_df = ts_df.astype(float)

display(ts_df)
```

	I1	I2	I3	t2	t3	UTS
0	330.0	310.0	1000.0	6.0	12.0	443.0
1	300.0	280.0	600.0	6.0	8.0	432.0
2	330.0	280.0	1000.0	4.0	12.0	445.0
3	330.0	310.0	1000.0	4.0	8.0	471.0
4	330.0	280.0	600.0	6.0	12.0	459.0
5	300.0	310.0	600.0	4.0	8.0	373.0
6	300.0	310.0	1000.0	6.0	8.0	463.0
7	330.0	280.0	1000.0	6.0	8.0	459.0
8	300.0	310.0	600.0	6.0	12.0	286.0
9	300.0	310.0	1000.0	4.0	12.0	460.0
10	300.0	280.0	1000.0	6.0	12.0	471.0
11	330.0	310.0	600.0	6.0	8.0	397.0
12	330.0	280.0	600.0	4.0	8.0	426.0
13	300.0	280.0	600.0	4.0	12.0	447.0
14	300.0	280.0	1000.0	4.0	8.0	455.0
15	330.0	310.0	600.0	4.0	12.0	358.0

Fig. 1.41 Importing MIAB_UTS

```
# Load the data and replace the '..' with nan
ts_df = pd.read_excel('MIAB_HD.xlsx')

ts_df = ts_df.astype(float)

display(ts_df)
```

	I1	I2	I3	t2	t3	HD
0	330.0	310.0	1000.0	6.0	12.0	194.0
1	300.0	280.0	600.0	6.0	8.0	173.0
2	330.0	280.0	1000.0	4.0	12.0	196.0
3	330.0	310.0	1000.0	4.0	8.0	203.0
4	330.0	280.0	600.0	6.0	12.0	191.0
5	300.0	310.0	600.0	4.0	8.0	170.0
6	300.0	310.0	1000.0	6.0	8.0	218.0
7	330.0	280.0	1000.0	6.0	8.0	212.0
8	300.0	310.0	600.0	6.0	12.0	158.0
9	300.0	310.0	1000.0	4.0	12.0	222.0
10	300.0	280.0	1000.0	6.0	12.0	207.0
11	330.0	310.0	600.0	6.0	8.0	166.0
12	330.0	280.0	600.0	4.0	8.0	183.0
13	300.0	280.0	600.0	4.0	12.0	176.0
14	300.0	280.0	1000.0	4.0	8.0	205.0
15	330.0	310.0	600.0	4.0	12.0	160.0

Fig. 1.42 Importing MIAB_HD

Fig. 1.43 Instruction to import seaborn and matplotlib library

```
import seaborn as sns
import pandas as pd
import matplotlib.pyplot as plt
```

```
# calculate the correlation matrix
corr = ts_df.corr()
# display the correlation matrix
display(corr)
# plot the correlation heatmap
sns.heatmap(corr, xticklabels=corr.columns, yticklabels=corr.columns, cmap='RdBu')
```

	l1	l2	l3	t2	t3	NTS
l1	1.000000	0.000000	0.000000	0.000000	0.000000	0.057394
l2	0.000000	1.000000	0.000000	0.000000	0.000000	-0.116767
l3	0.000000	0.000000	1.000000	0.000000	0.000000	0.874762
t2	0.000000	0.000000	0.000000	1.000000	0.000000	0.174161
t3	0.000000	0.000000	0.000000	0.000000	1.000000	0.114788
NTS	0.057394	-0.116767	0.874762	0.174161	0.114788	1.000000

```
<matplotlib.axes._subplots.AxesSubplot at 0x200b8867940>
```

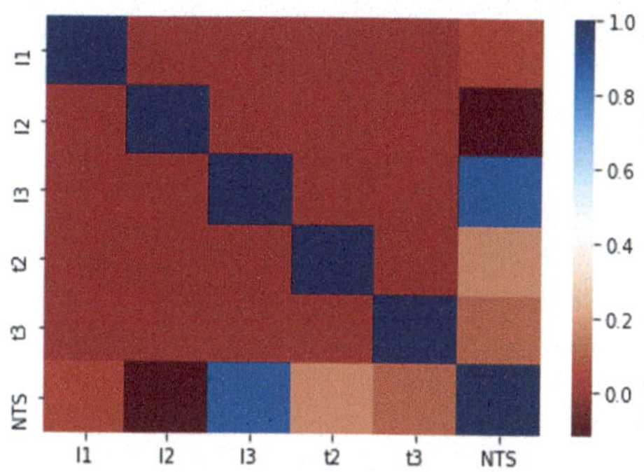

Fig. 1.44 MIAB_NTS co-relation matrix

1.5.4 Prediction Analysis Using Scikit-Learn

This section discusses contents pertaining to various linear models for regression and classification using scikit-learn. The algorithms which will be discussed in the section to give a clarity about predictive analysis are

Linear Regression

This algorithm consists of independent variables representing each feature of a dataset, dependent variable basically the output which is supposed to be predicted, intercept and coefficients of each feature represented in the form of the following equation (Fig. 1.50):

```
# calculate the correlation matrix
corr = ts_df.corr()
# display the correlation matrix
display(corr)
# plot the correlation heatmap
sns.heatmap(corr, xticklabels=corr.columns, yticklabels=corr.columns, cmap='RdBu')
```

	l1	l2	l3	t2	t3	UTS
l1	1.000000	0.000000	0.000000	0.000000	0.000000	0.090413
l2	0.000000	1.000000	0.000000	0.000000	0.000000	-0.436782
l3	0.000000	0.000000	1.000000	0.000000	0.000000	0.622701
t2	0.000000	0.000000	0.000000	1.000000	0.000000	-0.031835
t3	0.000000	0.000000	0.000000	0.000000	1.000000	-0.136256
UTS	0.090413	-0.436782	0.622701	-0.031835	-0.136256	1.000000

```
<matplotlib.axes._subplots.AxesSubplot at 0x17a1ede7f60>
```

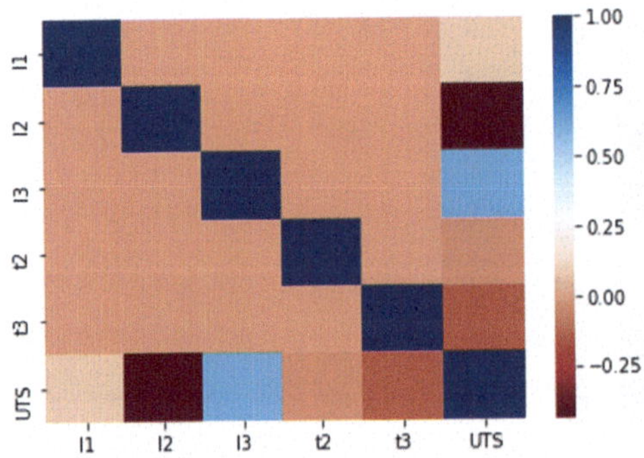

Fig. 1.45 MIAB_UTS co-relation matrix

$$\hat{y}(w, x) = w_0 + w_1x_1 + w_2x_2 \ldots + w_px_p$$

where

y	Dependent Variable
x	Independent Variable
w_0	Intercept
$w_1 \ldots w_p$	weight of each feature

```
# calculate the correlation matrix
corr = ts_df.corr()
# display the correlation matrix
display(corr)
# plot the correlation heatmap
sns.heatmap(corr, xticklabels=corr.columns, yticklabels=corr.columns, cmap='RdBu')
```

	l1	l2	l3	t2	t3	HD
l1	1.000000	0.000000	0.000000	0.000000	0.000000	-0.074652
l2	0.000000	1.000000	0.000000	0.000000	0.000000	-0.161747
l3	0.000000	0.000000	1.000000	0.000000	0.000000	0.870944
t2	0.000000	0.000000	0.000000	1.000000	0.000000	0.012442
t3	0.000000	0.000000	0.000000	0.000000	1.000000	-0.080873
HD	-0.074652	-0.161747	0.870944	0.012442	-0.080873	1.000000

```
<matplotlib.axes._subplots.AxesSubplot at 0x23b4be34a58>
```

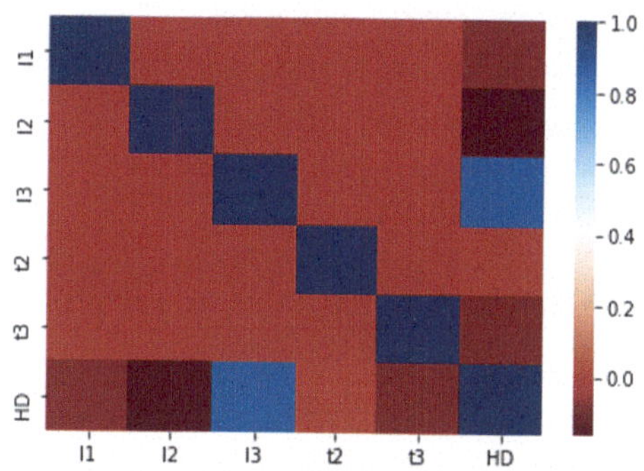

Fig. 1.46 MIAB_HD co-relation matrix

A standard procedure of splitting the data into 80% of training data and 20% of testing data from the dataset is followed to predict the dependent parameters across each table. It is a two-step process, in which both the steps are a common process to split the dataset into training set and testing set (Figs. 1.51, 1.52, 1.53 and 1.54).

The corresponding output of spitting each datasets are as follows.

```
# get the summary
desc_df = ts_df.describe()
# add the standard deviation metric
desc_df.loc['+3_std'] = desc_df.loc['mean'] + (desc_df.loc['std'] * 3)
desc_df.loc['-3_std'] = desc_df.loc['mean'] - (desc_df.loc['std'] * 3)
# display it
desc_df
```

	I1	I2	I3	t2	t3	NTS
count	16.000000	16.000000	16.000000	16.000000	16.000000	16.000000
mean	315.000000	295.000000	800.000000	5.000000	10.000000	646.250000
std	15.491933	15.491933	206.559112	1.032796	2.065591	65.231383
min	300.000000	280.000000	600.000000	4.000000	8.000000	553.000000
25%	300.000000	280.000000	600.000000	4.000000	8.000000	585.750000
50%	315.000000	295.000000	800.000000	5.000000	10.000000	642.000000
75%	330.000000	310.000000	1000.000000	6.000000	12.000000	709.250000
max	330.000000	310.000000	1000.000000	6.000000	12.000000	726.000000
+3_std	361.475800	341.475800	1419.677335	8.098387	16.196773	841.944149
-3_std	268.524200	248.524200	180.322665	1.901613	3.803227	450.555851

```
# filter the data frame to remove the values exceeding 3 standard deviations
econ_remove_df = ts_df[(np.abs(stats.zscore(ts_df)) < 3).all(axis=1)]
# what rows were removed
ts_df.index.difference(econ_remove_df.index)
```

```
Int64Index([], dtype='int64')
```

Fig. 1.47 MIAB_NTS data conditioning

Once the model is trained, the trained model is further employed to predict the data in the following manner by the help of the following instructions (Fig. 1.55).

The predictions for each dataset are observed as shown in Tables 1.11, 1.12 and 1.13.

From Fig. 1.56, it is observed that the actual value is quite in agreement with the predicted values of NTS.

```
# get the summary
desc_df = ts_df.describe()
# add the standard deviation metric
desc_df.loc['+3_std'] = desc_df.loc['mean'] + (desc_df.loc['std'] * 3)
desc_df.loc['-3_std'] = desc_df.loc['mean'] - (desc_df.loc['std'] * 3)
# display it
desc_df
```

	I1	I2	I3	t2	t3	UTS
count	16.000000	16.000000	16.000000	16.000000	16.000000	16.000000
mean	315.000000	295.000000	800.000000	5.000000	10.000000	427.812500
std	15.491933	15.491933	206.559112	1.032796	2.065591	50.690195
min	300.000000	280.000000	600.000000	4.000000	8.000000	286.000000
25%	300.000000	280.000000	600.000000	4.000000	8.000000	418.750000
50%	315.000000	295.000000	800.000000	5.000000	10.000000	446.000000
75%	330.000000	310.000000	1000.000000	6.000000	12.000000	459.250000
max	330.000000	310.000000	1000.000000	6.000000	12.000000	471.000000
+3_std	361.475800	341.475800	1419.677335	8.098387	16.196773	579.883084
-3_std	268.524200	248.524200	180.322665	1.901613	3.803227	275.741916

```
# filter the data frame to remove the values exceeding 3 standard deviations
econ_remove_df = ts_df[(np.abs(stats.zscore(ts_df)) < 3).all(axis=1)]
# what rows were removed
ts_df.index.difference(econ_remove_df.index)
```

```
Int64Index([], dtype='int64')
```

Fig. 1.48 MIAB_UTS data conditioning

From Fig. 1.57, it is observed that the actual value is quite in coherence with the predicted values of UTS.

From Fig. 1.58, it is observed that the actual value is quite in coherence with the predicted values of HD.

```
# get the summary
desc_df = ts_df.describe()
# add the standard deviation metric
desc_df.loc['+3_std'] = desc_df.loc['mean'] + (desc_df.loc['std'] * 3)
desc_df.loc['-3_std'] = desc_df.loc['mean'] - (desc_df.loc['std'] * 3)
# display it
desc_df
```

	I1	I2	I3	t2	t3	HD
count	16.000000	16.000000	16.000000	16.000000	16.000000	16.000000
mean	315.000000	295.000000	800.000000	5.000000	10.000000	189.625000
std	15.491933	15.491933	206.559112	1.032796	2.065591	20.752108
min	300.000000	280.000000	600.000000	4.000000	8.000000	158.000000
25%	300.000000	280.000000	600.000000	4.000000	8.000000	172.250000
50%	315.000000	295.000000	800.000000	5.000000	10.000000	192.500000
75%	330.000000	310.000000	1000.000000	6.000000	12.000000	205.500000
max	330.000000	310.000000	1000.000000	6.000000	12.000000	222.000000
+3_std	361.475800	341.475800	1419.677335	8.098387	16.196773	251.881325
-3_std	268.524200	248.524200	180.322665	1.901613	3.803227	127.368675

```
# filter the data frame to remove the values exceeding 3 standard deviations
econ_remove_df = ts_df[(np.abs(stats.zscore(ts_df)) < 3).all(axis=1)]
# what rows were removed
ts_df.index.difference(econ_remove_df.index)
```

```
Int64Index([], dtype='int64')
```

Fig. 1.49 MIAB_HD data conditioning

```
from sklearn.linear_model import LinearRegression
```

Fig. 1.50 Instruction to import linear regression model from scikit-learn library

```
# define our input variable (X) & output variable (Y)
X = ts_df.iloc[:, :-1]
Y = ts_df.iloc[:, 5]
# Split X and y into X_
X_train, X_test, y_train, y_test = train_test_split(X, Y, test_size=.20, random_state=0)
#X_train, X_test, y_train, y_test = model_selection.train_test_split(X, y, train_size=0.75,test_size=0.25, random_state=101)
print ("X_train: ", X_train)
print ("y_train: ", y_train)
print("X_test: ", X_test)
print ("y_test: ", y_test)

# create a Linear Regression model object
regression_model = LinearRegression()

# pass through the X_train & y_train data set
regression_model.fit(X_train, y_train)
```

Fig. 1.51 Instructions for splitting and training the model

Fig. 1.52 Output from MIAB_NTS split instructions

```
X_train:            I1      I2       I3   t2     t3
13  300.0  280.0   600.0  4.0  12.0
4   330.0  280.0   600.0  6.0  12.0
2   330.0  280.0  1000.0  4.0  12.0
14  300.0  280.0  1000.0  4.0   8.0
10  300.0  280.0  1000.0  6.0  12.0
7   330.0  280.0  1000.0  6.0   8.0
15  330.0  310.0   600.0  4.0  12.0
11  330.0  310.0   600.0  6.0   8.0
3   330.0  310.0  1000.0  4.0   8.0
0   330.0  310.0  1000.0  6.0  12.0
5   300.0  310.0   600.0  4.0   8.0
12  330.0  280.0   600.0  4.0   8.0
y_train:  13    634.0
4      619.0
2      705.0
14     675.0
10     726.0
7      717.0
15     557.0
11     642.0
3      642.0
0      726.0
5      570.0
12     591.0
Name: NTS, dtype: float64
X_test:           I1      I2       I3   t2     t3
1  300.0  280.0   600.0  6.0   8.0
6  300.0  310.0  1000.0  6.0   8.0
8  300.0  310.0   600.0  6.0  12.0
9  300.0  310.0  1000.0  4.0  12.0
y_test:  1    562.0
6     713.0
8     553.0
9     708.0
Name: NTS, dtype: float64
```

Fig. 1.53 Output from MIAB_UTS split instructions

```
X_train:            I1        I2          I3    t2       t3
13  300.0  280.0    600.0  4.0  12.0
4   330.0  280.0    600.0  6.0  12.0
2   330.0  280.0  1000.0  4.0  12.0
14  300.0  280.0  1000.0  4.0   8.0
10  300.0  280.0  1000.0  6.0  12.0
7   330.0  280.0  1000.0  6.0   8.0
15  330.0  310.0    600.0  4.0  12.0
11  330.0  310.0    600.0  6.0   8.0
3   330.0  310.0  1000.0  4.0   8.0
0   330.0  310.0  1000.0  6.0  12.0
5   300.0  310.0    600.0  4.0   8.0
12  330.0  280.0    600.0  4.0   8.0
y_train:  13     447.0
4      459.0
2      445.0
14     455.0
10     471.0
7      459.0
15     358.0
11     397.0
3      471.0
0      443.0
5      373.0
12     426.0
Name: UTS, dtype: float64
X_test:           I1        I2          I3    t2       t3
1  300.0  280.0    600.0  6.0   8.0
6  300.0  310.0  1000.0  6.0   8.0
8  300.0  310.0    600.0  6.0  12.0
9  300.0  310.0  1000.0  4.0  12.0
y_test:  1    432.0
6    463.0
8    286.0
9    460.0
Name: UTS, dtype: float64
```

```
X_train:           I1      I2       I3    t2     t3
13  300.0  280.0   600.0  4.0  12.0
4   330.0  280.0   600.0  6.0  12.0
2   330.0  280.0  1000.0  4.0  12.0
14  300.0  280.0  1000.0  4.0   8.0
10  300.0  280.0  1000.0  6.0  12.0
7   330.0  280.0  1000.0  6.0   8.0
15  330.0  310.0   600.0  4.0  12.0
11  330.0  310.0   600.0  6.0   8.0
3   330.0  310.0  1000.0  4.0   8.0
0   330.0  310.0  1000.0  6.0  12.0
5   300.0  310.0   600.0  4.0   8.0
12  330.0  280.0   600.0  4.0   8.0
y_train:  13    176.0
4      191.0
2      196.0
14     205.0
10     207.0
7      212.0
15     160.0
11     166.0
3      203.0
0      194.0
5      170.0
12     183.0
Name: HD, dtype: float64
X_test:           I1      I2       I3    t2     t3
1  300.0  280.0   600.0  6.0   8.0
6  300.0  310.0  1000.0  6.0   8.0
8  300.0  310.0   600.0  6.0  12.0
9  300.0  310.0  1000.0  4.0  12.0
y_test:  1    173.0
6     218.0
8     158.0
9     222.0
Name: HD, dtype: float64
```

Fig. 1.54 Output from MIAB_HD split instructions

```
# Get multiple predictions
y_predict = regression_model.predict(X)

# Show the predicions
y_predict[:]

df = pd.DataFrame({'Actual': Y, 'Predicted': y_predict})
df
```

Fig. 1.55 Instructions to predict by using the trained model

Table 1.11 Results of NTS predicted data

NTS	NTS_Predicted
726	713.79
562	598.58
705	704.3125
642	658
619	629.458333
570	566.1875
713	707.91
717	720.6875
553	598.68
708	680.54
726	727.9375
642	606.43
591	592.958333
634	612.208
675	679.43
557	579.0625

Table 1.12 Results of UTS predicted data

UTS	UTS_Predicted
443	441.12
432	427.375
445	459.12
471	463.62
459	450.875
373	379.625
463	453.625
459	469.375
286	292.125
460	452.375
471	467.125
397	391.375
426	431.375
447	439.125
455	464.625
358	365.125

Table 1.13 Results of HD predicted data

HD	HD_Predicted
194	193.7375
173	176.15
196	202.1375
203	194.0625
191	182.1875
170	168.025
218	210.7
212	213.1875
158	167.7
222	215.65
207	206.775
166	174.1125
183	182.5125
176	176.1
205	207.1
160	163.0625

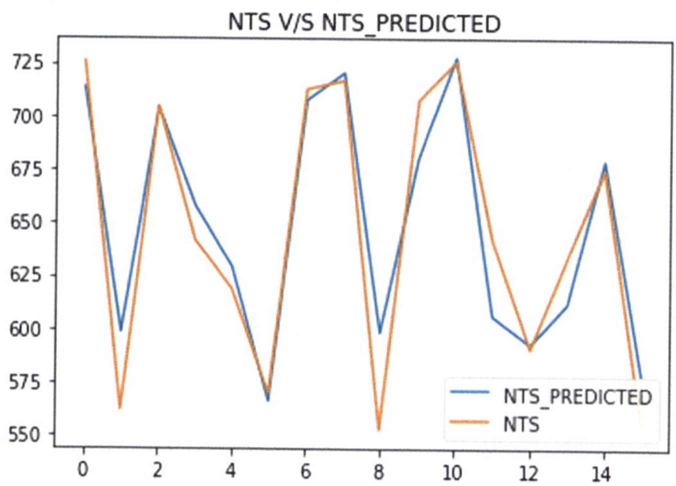

Fig. 1.56 NTS V/S NTS_predicted

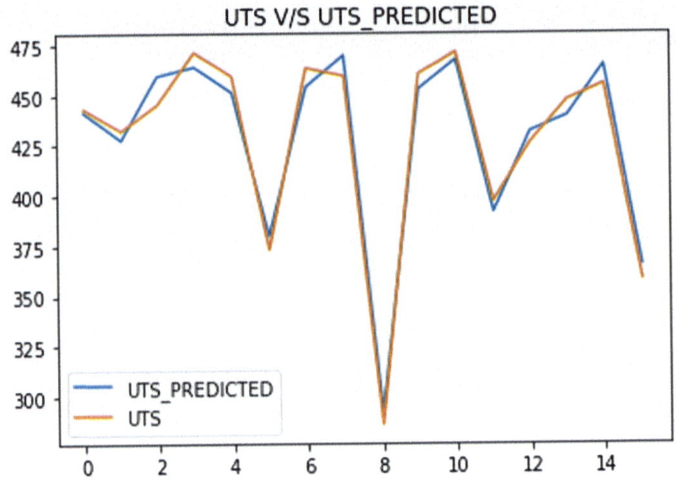

Fig. 1.57 UTS V/S UTS_predicted

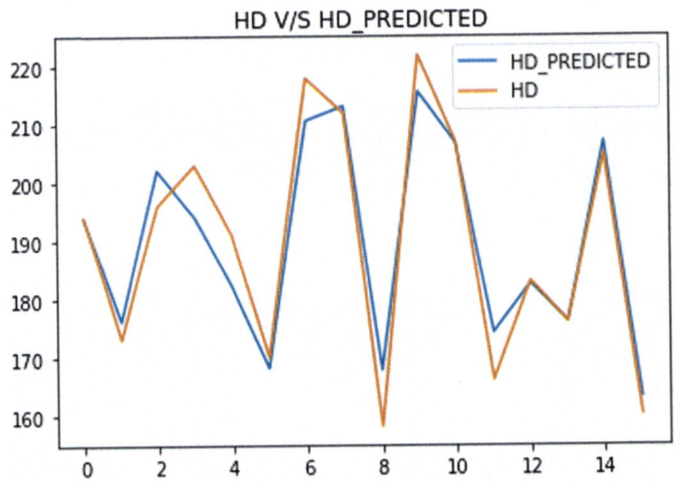

Fig. 1.58 HD V/S HD predicted

References

1. Messler Jr, Robert W (2008) Principles of welding: processes, physics, chemistry, and metallurgy. John Wiley & Sons
2. Kachinskiy VS, Krivenko VG (2002) Magnetically impelled arc butt welding of hollow and solid parts. Weld World 46(7/8):49–56
3. Kim JW, Choi DH (2003) A study on the numerical analysis of magnetic flux density by a solenoid for magnetically impelled arc butt welding. Proc Inst Mech Eng Part B J Eng Manuf 217(10):1401-1407
4. Welz Nentwig (1988) Effects of shielding gas, of the heating and upsetting process in pressure welding using a magnetically moved arc. Weld Cut 4:E65–E68
5. Ganowski FJ (1974) The magnetarc welding process. Weld Metal Fabric 206–213
6. Géron A (2019) Hands-on machine learning with Scikit-Learn, Keras, and TensorFlow: concepts, tools, and techniques to build intelligent systems. O'Reilly Media
7. Steffen Welz (1982) Pressure welding of tubes with a magnetically displaced arc. Schweißen und Schneiden Translation 4:E70–E72
8. Nentwid Appel (1995) Coil systems, magnetic field distribution and arc behaviour in pressure welding with magnetically impelled arc. Weld Cut 1:E9–E11
9. Kuchuk-Yatsenko SI et al (1988) Control of the arc moving in a narrow gap under the effect of a magnetic field in press welding of pipes. Weld Int 11:965–968
10. Sato T, Katayama J, Ioka S, Otani M (1991) An experimental study of rotational behaviour of the arc during magnetically impelled arc butt welding. Weld Int 5(1):5–10
11. Kuchuk-Yatsenko SI, Syrovatka V, Kuznetsov P (1983) Speed of movement by an arc in the gap between a tube and a plate under the action of a magnetic field. Autom Weld 23–26
12. Taneko A, Arakida F, Takagi K (1987) Analysis of arc rotation velocity in magnetically impelled arc butt welding. Weld Int 1(3):247–253
13. Xiancong H, Ruilin X et al (1984) A study of magnetically impelled arc butt welding. The Welding Institution of The Chinese Mechanical Engineering Society, 6–8 September 1984, pp 1–6
14. Kalev L, Mikhailov VG, V'lchev GP Calculation of temperature fields in welding with a magnetically controlled arc with the power changing with time. Weld Prod 28–29 (1986)
15. Georgescu V, Iordachescu D, Scutelnicu E Study of magnetizing systems for magnetarc welding. (Other Details Unavailable)
16. Edson DA (1982) Magnetically impelled arc butt (MIAB) welding of thicker walled tubes. (Other Details Unavailable)
17. Vendan SA, Manoharan S, Buvanashekaran G, Nagamani C Development of a MIAB welding module and experimental analysis of rotational behavior of arc—simulation of electromagnetic force distribution during MIAB welding of steel pipes using finite element analysis. Int J Adv Manuf Technol 43(11–12):1144–1156 (2009)
18. Fletcher L, Stecher G, Stubbs G (2005) Developments in the application of MIAB welding to oil and gas pipelines. In Proceedings of International conference on pipeline construction technology, pp 1–11
19. Schlebeck E (1978) Welding with a magnetically moved arc (MBL welding): a new means of rationalization. In: Proceedings of welding institute conference on advances in welding processes. Harrogate, pp 249–256
20. Westgate SA, Edson DA (1986) The MIAB welding of tubular sections for mass production industries (No. 860582). SAE Technical Paper
21. Hagan D (1979) An industrial application of MIAB welding of tubes: a rear axle cross tube assembly. In: Second international conference on pipe welding, London, 20–22 November 1979, vol 1, Paper 17, pp 51–58
22. Hiller F, Schmidt M, Breiksch J Use of the magnetarc welding process in the production of truck cab suspension systems. In: ThyssenKrupp Techforum, December 2003, pp 40–43

23. Jenicek A, Cramer H (2005) Bush weld on aluminium materials—a further development for joining of small hollow bodies using a magnetically impelled arc. Weld Cut 4(3):4–10
24. Mori Yasuda (1990) Magnetically impelled arc butt welding of aluminum pipes. Trans Jpn Weld Soc 21(1):3–10
25. Vendan SA, Manoharan S, Buvanashekaran G, Nagamani C Magnetically Impelled Arc Butt Welding of alloy steel tubes in boilers–establishment of parameter window. Mechatronics 21 (1):30–37 (2011)
26. McKinney W (2012) Python for data analysis: data wrangling with Pandas, NumPy, and IPython. O'Reilly Media Inc

Chapter 2
Supervised Machine Learning in Cold Metal Transfer (CMT)

2.1 Introduction

The cold metal transfer process evolved from the MIG/MAG process to address the lacunas posed during the aluminium and other materials joining. CMT being controlled process permits the material transfer with the nominal flow of current thereby reducing power consumption and material wastage. Constant retraction of filler wire at very short intervals is one of the major sequential steps in CMT process. CMT welding is patented by the Fronius International (Pettenbach, Austria). Cold metal transfer welding is exclusively designed for the metals where intermetallic compounds are serious challenge. Higher the heats produced the greater the degree of intermetallic formation. CMT which does not melt the base metal completely, instead it wet the surface and braze the joining at the interface. Combination of brazing and short circuit mode metal transfer makes efficient material transfer to weld without spattering. CMT process is fully digitalized with remote control unit (RCU) to control the welding parameters.

2.2 Principle of System Operation

Cold metal transfer welding is a sophisticated version of GMAW which incorporates filler wire feed drive into closed-loop feedback control. This advanced joining method yields extremely stable arc, incorporating filler wire feed drive, ease of joining thin metal sheets, lower joule heat input and sprinkle free molten metal transfer facilitates the exploration of several combinations of material joining as shown in Table 2.1. Cold metal transfer (CMT) welding enables controlled heat input feature. CMT decouples the electric arc transient behaviour from the rate of filler wire feed. It features includes the remote control unit (RCU) makes the process control fully digitalized. The wire feed rate in the CMT process is

S. A. Vendan et al., *Welding and Cutting Case Studies with Supervised Machine Learning*, Engineering Applications of Computational Methods 1,
https://doi.org/10.1007/978-981-13-9382-2_2

Table 2.1 Cold metal transfer features [1]

Wire feed drive	The wire retraction movement when detects short circuit helps to drop the metal globules up to 70 t/s
Spatter-free	The backward movement of the filler wire helps in detaching the globules during short circuit mode. In short circuit mode, current is negligible and the spatter-free metal transfer takes place
Lower heat input	With closed-loop feedback control of power source and filler wire actuator, the metal transfer takes place with almost no electrical input, which in turn controls the intermetallic layer width
Stable arc	The electric arc in cold metal transfer is driven from the pulse modulated—power source which almost operates independently with the nature of base metal to be welded and speed of the welding process

controlled by a microprocessor-based controllers, which operates the feed actuators and these microcontrollers ensure the rate of filler wire feeding almost independent on the electrical characteristics of power source. Short circuit metal transfer is main mode of material transfer in CMT. The metals (filler and base metals) can be melted by the energy supplied by controlled varying of the arc cycle phase and the rate of wire feed.

The basic principle of CMT process is to operate in the short circuit mode (dip transfer) characterized by the low current and voltage, i.e. low heat input. An initial high pulse of current is supplied, forming an arc between the advancing electrode wire and the substrate, which melts the tip of electrode and the part of base metal. The current is reduced almost to zero magnitude as soon as filler wire touches base material (short circuit occurs) and the molten droplets at the end of the filler wire are dropped into molten liquid pool of base metal in short circuit mode. After the droplet detachment from filler wire, the wire retracts back to its original position and the process repeats on. Source: Pickin et al. [2]. Figure 2.1 shows the images of cold metal transfer advanced working principle taken by magnified lens.

The term "cold" is relative. With the comparison to existing welding process, the CMT is colder. The heat utilized is almost 30% of MIG/MAG welding process. The key features of CMT are presented in Table 2.1.

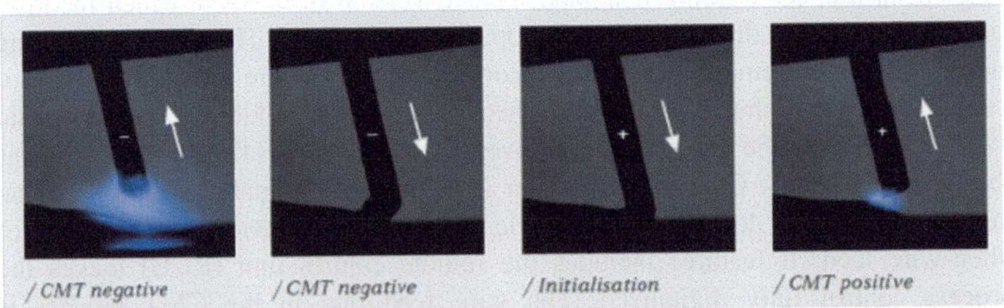

Fig. 2.1 Graphical view of cold metal transfer advanced process operation [1]

2.2.1 Welding Power Sources

The wire feed is integrated into the welding torch which requires maintaining wire feed rate proportional to the rate of consumption of wire during arcing period. Of phase of these actions, result in unstable arc with high voltages which in turn produces undesirable heat input and improper welds. The relationship between welding voltage and current (dynamic characteristics of the power source) is to be cautiously chosen in accordance with the welding process to evade non linearity. If the consumption rate of filler metal does not follow the feeding rate of the wire, even small change in arc length gives enormous rise in welding voltage and current, as observed in the constant voltage/flat characteristic power sources. Inverse of this is drooping characteristics/constant power as shown in Fig. 2.2, where welding current results in small change with the increase in welding voltage while the heat input remains almost constant. This characteristic is highly preferred in welding of aluminium and its alloys.

Figure 2.3 shows the CMT power source output waveform. The waveform is mainly divided into two cycles: (1) Arcing period (2) Short circuit period. As described previously in the process principle section, the CMT operates in short circuit mode for metal transfer with no heat input with negligible voltage and current (Fig. 2.3) is drawn from the source. The arcing phase melts both filler and base metals. Wherein, high current pulses provided to the filler wire increases the rate of consumption of filler wire. Further, the high heat input leads to the melting of the metals and transfer of material in the short circuit mode.

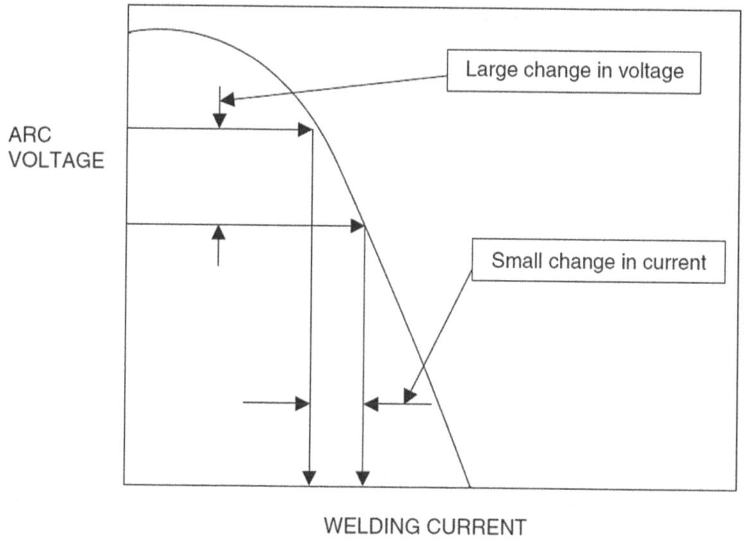

Fig. 2.2 Drooping characteristic of powers source [1]

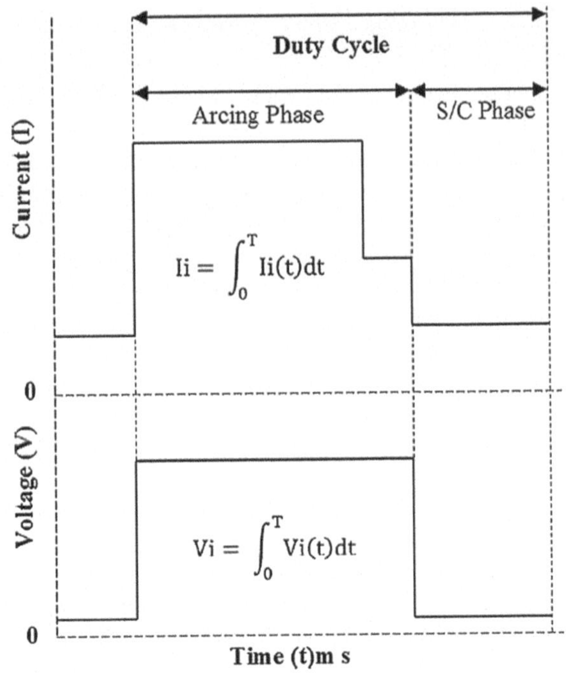

Fig. 2.3 CMT power source instantaneous voltage and current cycle waveforms. *Source* Pickin et al. [2]

2.2.2 *Different Modes of Operation in CMT*

CMT process comprises of four key operating modes:

1. CMT
2. CMT pulse
3. CMT advanced
4. CMT advanced pulse

The features and advantages of different modes of CMT are briefly shown in Table 2.2.

2.3 Process-State of Art

The basic operating principles of the CMT process were previously reported by Pickin and Young [3] and Jicai Feng (2013). Pickin and Young [3] performed experiments with CMT and MIG for a comparative evaluation. The results revealed that penetration depth, deposition rate, applicability for thicker materials and reduced heat input were good in CMT process. The advantages of low dilution

Table 2.2 Features of different modes of operation in CMT

CMT process	CMT + PULSE	CMT advanced	CMT advanced pulse
1. Wire movement incorporated into process control 2. Controlled, low heat input 3. Extremely stable arc 4. Economical due to higher welding speeds	1. A combination of pulsed cycles with CMT cycles 2. Increased heat input 3. Higher welding speed 4. Specific and variable addition of pulses 5. Widespread performance and flexibility	1. Polarity of welding current is incorporated into process control 2. Change of polarity takes place in short circuit mode 3. Heat input is controlled and precise to the weld bead. 4. Gap-bridge ability is high due to spatter-free material transfer	1. Combination of electrode negative CMT and electrode positive pulsed arc cycles 2. Absolute precision 3. Greatest mastery of the arc

cladding over Aluminium alloy base metal plates using CMT are studied by Pickin et al. [2] and Rajeev et al. (2012).

The introduction of the concept for cold metal joining has provided ample scope for its feasible employment in joining dissimilar metals. Some of the prominent literature directly contributing to this study are briefly presented in Table 2.3. These studies have generally focused upon the applicability of CMT process in joining of dissimilar metals and highlighted the advantage of CMT process advantage over the existing fusion welding processes.

Baoqiang Cong et al. [13] discussed the effects of various arc modes adopted by CMT namely conventional CMT process, CMT pulse (CMT-P), CMT advanced (CMT-ADV) and CMT pulse advanced (CMT-PADV) performed on additively

Table 2.3 Briefing on literature survey

Author name	Reported information
Zhang et al. [4]	Joining aluminium to zinc-coated steel
Cao et al. [5–10]	Joining of aluminium-to-galvanized mild steel Joining of magnesium AZ31B-to-aluminium A6061-T6 Microstructural and mechanical properties analysis of welding–brazing of titanium to copper Cold metal transfer spot plug welding of AA6061-T6-to-galvanized steel Welding–brazing of pure titanium TA2 to magnesium alloy AZ31B
Minjung Kang et al. [11]	Joining of Al 5052 alloy to aluminized steel sheet
Beytullah Gungor et al. [12]	Mechanical and microstructural properties studies on 5083-H111-to-6082-T651 aluminium alloys weldments

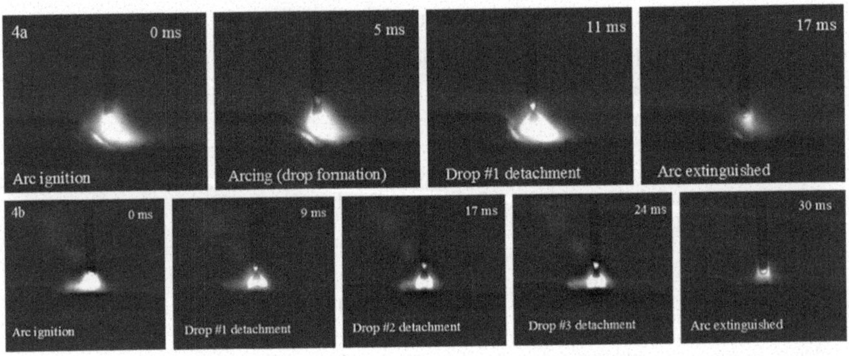

Fig. 2.4 Arcing phase droplet detachment *Source* Pickin et al. [2]

manufactured Al-6.3% Cu alloy for porosity characterization. Arc phase droplet detachment is shown in Fig. 2.4.

2.4 Experimentation and Process Governing Parameters

2.4.1 Materials

The material chosen for this research study is Aluminium alloy 6061 (AA6061-plate (Fig. 2.5)) owing to its exceptional physical (low weight, high strength) and chemical (corrosion resistance, non-toxicity) properties which finds prominent applications n sectors, viz., aerospace, automobile, ship building, construction and food packaging. The chemical composition is shown in Table 2.4.

(a) Physical properties
Density: 2.7 g/cm^3
Melting Point: 582 °C (approx.)
Modulus of elasticity: 70–80 GPa
Poissons Ratio: 0.33
(b) Thermal Properties
Thermal conductivity: 173 W/m K
Coefficient of thermal expansion: 25.2×10^{-6} m/m °C
(c) Electrical properties
Electrical resistivity: 3.99×10^{-6} Ω-cm

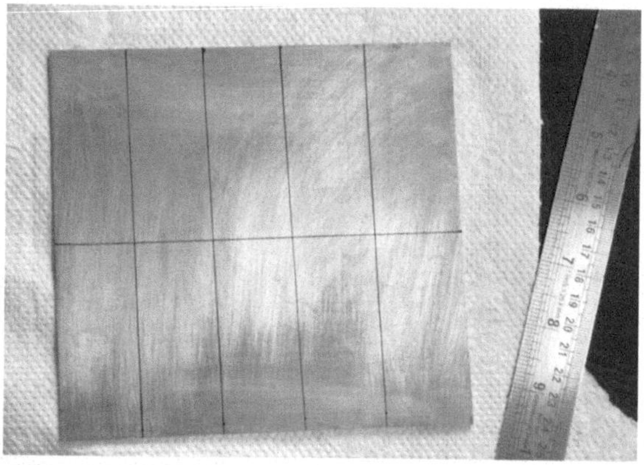

Fig. 2.5 AA6061 plates used for welding [1]

Table 2.4 Nominal chemical composition of Aluminium alloy 6061 alloy plates [1]

Component	Al	Mg	Si	Fe	Cu	Zn	Ti	Mu	Cr	Others
Amoun (wt%)	Bal.	0.8–1.2	0.4–0.8	0.7	0.15–0.40	0.25	0.15	0.15	0.04	0.05

2.5 Experimentation

2.5.1 CMT Welding Set-up

Welds were made on the Aluminium 6061 alloy plate by using a Fronius advanced CMT 7000 VR machine available at NIT Trichy, India. Welding trials are performed by employing Pulse-CMT mode normally produces larger variations in contact angles when compared with CMT mode. The variable speed of welding torch is achieved by a 6-axis Yaskawa Motoman robot available at NIT Trichy, India. Welding speed and path can be varied by robotic arm and the other parameters are controlled by the remote control unit (RCU) of the CMT machine. The weld set-up available with NIT Trichy, India is employed for the study (Fig. 2.6). The argon cylinder used in this study is configured to release the argon gas at a fixed flow rate of 18 l/min. further, the shielding is made void-free to avoid leakage and interference from external atmospheric impurities.

2.5.2 Welding Process Parameters

Literature survey emphasizes that the parameters welding current and speed have significant effects on the weldment quality and strength of the metals. In case of welding aluminium, the heat input supplied to the weld metal should be less in

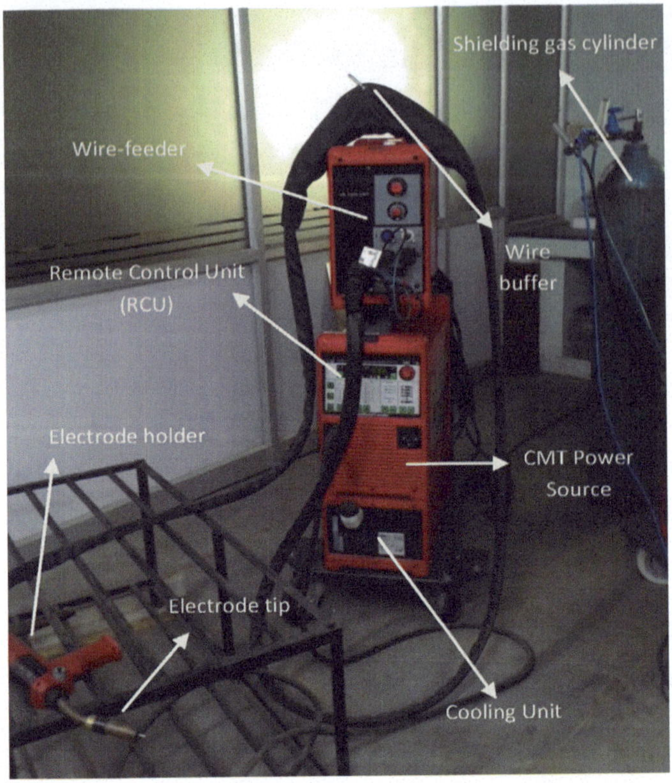

Fig. 2.6 CMT 7000 VR machine with parts pointed out [1]

Table 2.5 Selection of process parameters

Process parameters
Welding current
Welding speed
Fixed parameters
Arc length
Wire feed rate
Shielding gas flow rate
Electrode material

magnitude which is proportional to the welding current. In manufacturing sector, speed is one of the key parameters during selection for increased productivity. The fixed and variable process parameters for CMT are presented in Table 2.5.

In this research study, welding current and welding speed is varied during experimentation by fixing other parameters constant for effective analysis of process parameters.

2.5.3 Estimation of Heat Input Calculations

Non waveform power supplies heat input equation

$$\text{Heat input}\, HI\, (\text{J/mm}) = \frac{Voltage \times Amperage \times 60}{Travel\, speed\, (\text{mm/min})} \tag{1}$$

In CMT welding, the one cycle time period (T) of waveform is used for (i). Time taken for peak pulse period (t_1) constitutes 50% of T (ii). Time taken for base period (t_2) and short circuit phase (t_3), each constitutes 25% of T.

According to ASME Section IX QW-409.1, heat input equation for waveform controlled power supplies is

$$HI\, (\text{J/mm}) = \frac{Power\, (\text{J/s}) \times Arctime\, (\text{s})}{Weldbead - length\, (\text{mm})} \tag{2}$$

Therefore, total time taken in welding process is expressed as

$$t = t_1 + t_2 + t_3 \tag{3}$$

Arc time is divided into three intervals. t_1—peak current interval, t_2—base current interval, t_3—short circuit current interval, where V is almost zero, leading to trivial power flow.

Total power supplied per cycle by the supply is

$$\begin{gathered} VI_p\frac{t}{2} + VI_b\frac{t}{4} + VI_{sc}\frac{t}{4} \\ Vt\left(\frac{I_p}{2} + \frac{I_b}{4} + \frac{I_{sc}}{4}\right) \end{gathered} \tag{4}$$

where I_p, I_b, I_{sc} denotes peak current, base current, short circuit current.

Heat input equation adopted for calculations are [1]

$$HI_{CMT} = Vt\left(\frac{I_p}{2} + \frac{I_b}{4} + \frac{I_{sc}}{4}\right) \Big/ Weldbead - length \tag{5}$$

CMT illustrates greater efficiency as compared to the non-periodic, i.e. continuous power supplies used in conventional welding. The integrated wire feed motion allows the controllable detachment of molten metal from filler wire with almost no current flowing in the circuit. This permits a lower heat input to the base metal which is essential for controlling the formation of intermetallic phases. The efficiency rises by 25% with the application of CMT welding. To justify the efficiency validation, identical values of welding current, welding voltage and welding speed are chosen.

2.5.4 Results and Discussions

2.5.4.1 Metallurgical Analysis

The preparation of samples for evaluating the metallurgical integrity of the material with respect to electrical parameters was prepared according to the American standard for materials and testing (ASTM) E3 standardized metallographic technique.

2.5.4.2 Macrostructural Examination

After polishing and etching, from the physical inspection of sectioned samples, one can observe the heat-affected zone in base metal and filler metal composition through the un-aided eye. Macrostructural images were used to analyse the variations in weld bead shape. The CMT weldments are shown in Fig. 2.7. The starting

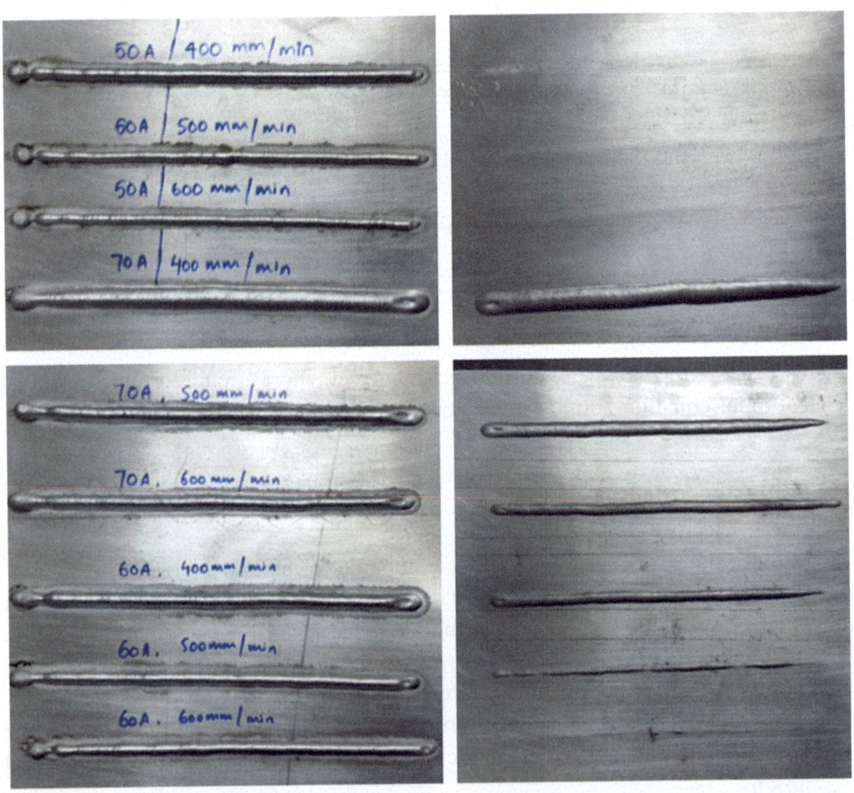

Fig. 2.7 CMT welded samples of AA6061 [1]

phase of arc in welding Al alloys requires extra ignition to prevent fusion defects. Owing to this, the base metal has to start melting right away with peak current ignition of arc. With the increase in current, the weld has remarkable depth of penetration.

From Fig. 2.7 for higher currents, imaging scars/surface modifications are visible on the rear side which contradicts the observations for lower currents. From the visible inspection of welded specimens, one can observe ripple profile around the weld bead. Degree of ripple appearance becomes clearer with increasing welding speed. Occurrence of ripple profile is dependent on welding parameters, filler wire composition, etc. Degree of ripples is significantly governed by the welding speed by keeping other parameters fixed. Higher welding speeds results in less heat input per unit length and the filler metal with less fluidity tends to form the high ripple content in the weld profile. With welding speed of 10 mm/s, higher ripples. While low welding speeds leads to better fluidity of weld pool and smooth ripple can be observed.

2.5.4.3 Weld Bead Geometry Measurements

The measurements of weld bead geometry were made from images of the macrostructures.

The geometry of weldment was made from the images of the welded specimens. The weld bead geometry comprises the following:

1. Depth of penetration (DOP),
2. Reinforcement height (RH),
3. Weld pool width (WW),
4. Weld penetration area (Ap), and
5. The weld reinforcement area (Ar).

The image analysis software IMAJE J software (open source) was used for the measurements.

2.5.4.4 Effect of Welding Current and Welding Speed on the Geometry of Weld Bead

Analysis of geometry of clads in terms of depth of penetration (mm), weld pool width (mm) and reinforcement height (mm) (Fig. 2.7) is performed. Weld bead geometry characteristics are obtained by usage of optical microscopy (OM) for each sample (Table 2.6) for various trails of welding current (Iw) and welding speed (Sw) parameters. Weld geometry properties express with respect to Iw and Sw individually are presented in Table 2.7. The following observations are made from the experimental analysis:

Table 2.6 CMT welded specimens (10× Magnification) at different input process parameters [1]

Welding current (A)	Welding speed (mm/min)	Weld bead appearance	
		Top side	Bottom side
50	400	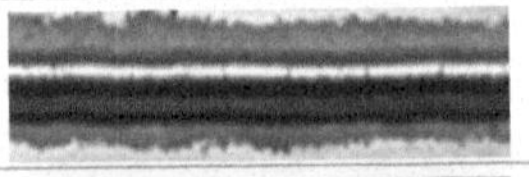	No penetration
50	500	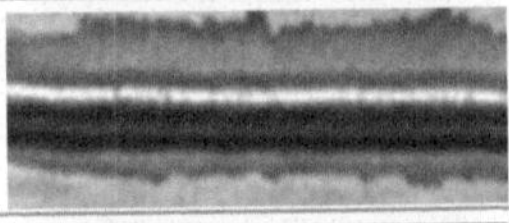	No penetration
50	600	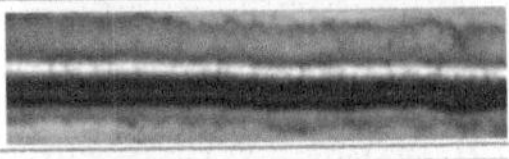	No penetration
60	400	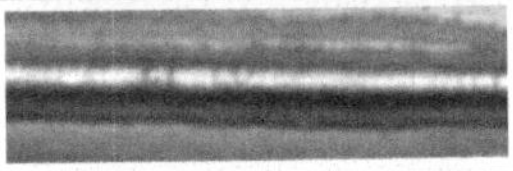	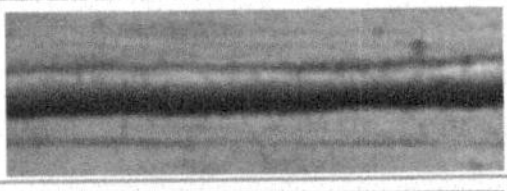
60	500	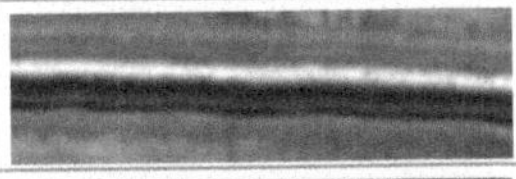	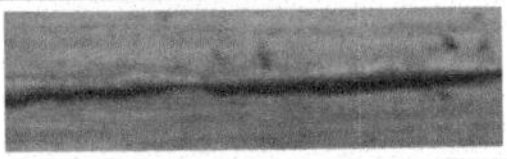
60	600	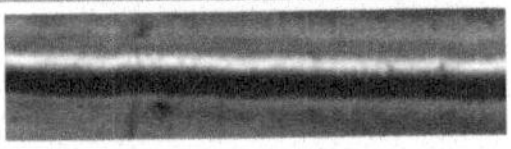	
70	400	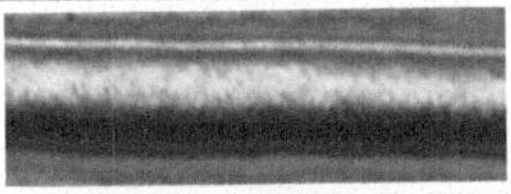	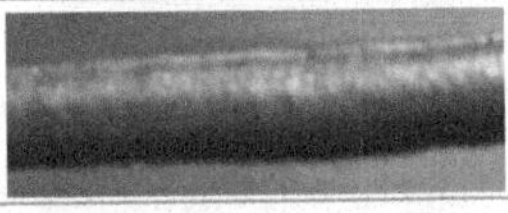
70	500		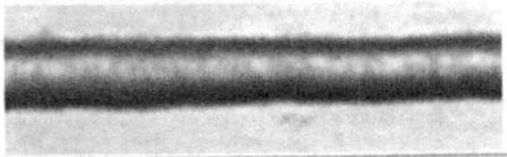
70	600		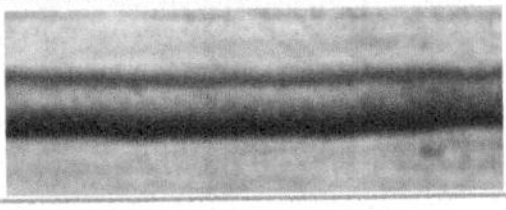

Table 2.7 Welding parameters and weld geometry of the samples [1]

S. no.	Welding current (A)	Welding speed (mm/min)	Depth of penetration (mm)	Weld pool width (mm)	Reinforcement height (mm)
A	50	400	0.51	3.22	2.69
B	50	500	0.33	2.32	2.53
C	50	600	0.32	1.75	2.51
D	60	400	2.68	5.49	2.55
E	60	500	2.34	4.89	2.5
F	60	600	1.67	4.37	2.64
G	70	400	2.98	6.8	2.14
H	70	500	2.56	5.3	1.93
I	70	600	2.63	5.61	2.06

- Welding current is directly proportional to the weld pool width. As current increases, force on the molten metal drops which lead to more spattering of metal. This decreases with increase in welding speed while maintaining constant welding current. This contributes lesser amount of deposited material per unit length and low driving force for molten weld pool to cover on the surface of the base metal.
- Depth of penetration is governed by arc force, penetration force of droplet and heat input supplied that is proportional to Iw (due to increased centrifugal and Coulomb's forces). With increase in Sw, weld depths decrease due to lower heat inputs.
- The reinforcement height decreases with increase in current due to larger expansion area and penetration force of molten metal pool, which is also slightly decreased with increased welding speeds.

2.5.4.5 Porosity Characteristics

Pores caused by hydrogen are prominently observed while welding aluminium. Liquid aluminium has an affinity to absorb large amount of hydrogen during solidification which leads to occurrence of pores in the weldments. During solidification, the hydrogen solubility decreases, thus the excess gas is released and the hydrogen bubbles get trapped and forms pores. Lower welding speeds provide higher heat input per area leading to a wider fusion zone. As a result of lower solidification rate, declines the porosity distribution. When the higher welding currents are applied, the high solubility of molten aluminium for hydrogen tends to absorb additional hydrogen in the molten metal pool during welding. However, in CMT welding, the confined heat input with short circuit transfer of metal in shielding gas reduces the formation of pores considerably. From Table 2.8, it may be observed that the pores are lesser at lower welding speeds and higher welding currents.

After setting the current and feed rate, nine weld beads were prepared at different welding speeds of 6.6 mm/s, 8.3 mm/s, and 10 mm/s (corresponding to 400 mm/min, 500 mm/min, and 600 mm/min, respectively). The images of polished cross section of nine beads are shown in Table 2.8. The first thing to note is that the cross-sectional area of the bead and, hence, the volume of the bead per unit length reduces with increasing welding speed. Heat per unit length reduces with increasing welding speed. This quantity is calculated by dividing the electrical power input by the welding speed, and a factor is introduced to take care of the conversion efficiency. It is obvious that higher the thermal heat input per unit length, the higher is the dilution. The dilution is calculated from the percentage of the total area of cross section of the bead, which gets depicted on the surface of the substrate as shown in Fig. 2.8. In another way, dilution can be obtained for each of the major elements in filler wire and is obtained by the following formula:

Table 2.8 Macrographs of weldments [1]

Macroscopic view of samples cross section	Iw (A)	Sw(mm/min)	Remarks
	50	400	• Insufficient depth of penetration of filler metal • Lower area of fusion zone • Weld bead contact angle >90° • Porosity density is low
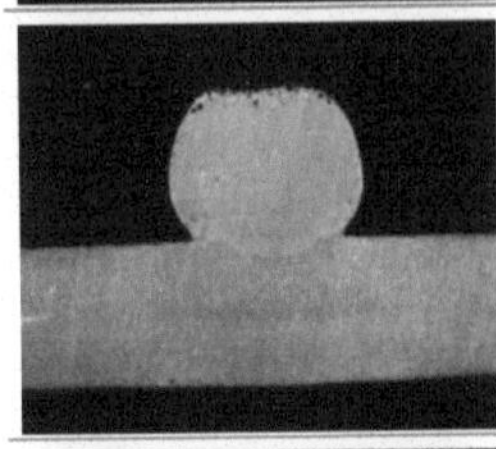	50	500	• Insufficient depth of penetration of filler metal • Lower area of fusion zone • Weld bead contact angle >90° • Porosity density is low
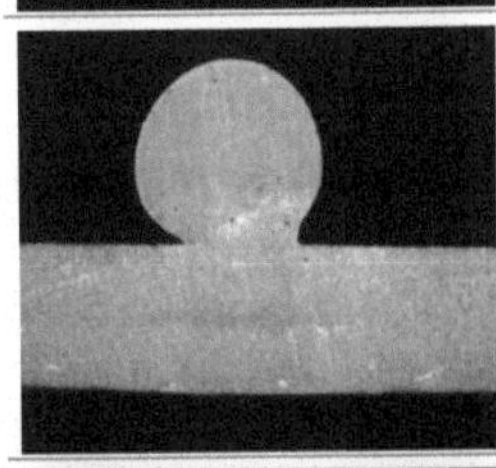	50	600	• Insufficient depth of penetration of filler metal • Low area of fusion zone • Weld bead contact angle >90° • Porosity density is low
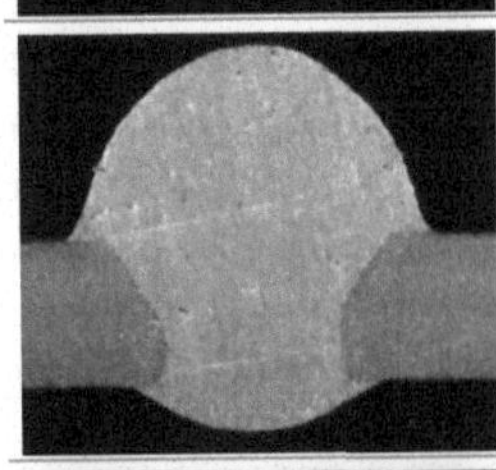	60	400	• Sufficient depth of penetration of filler metal • Larger area of fusion zone • Weld bead contact angle $\leq 90°$ • Porosity density is moderate
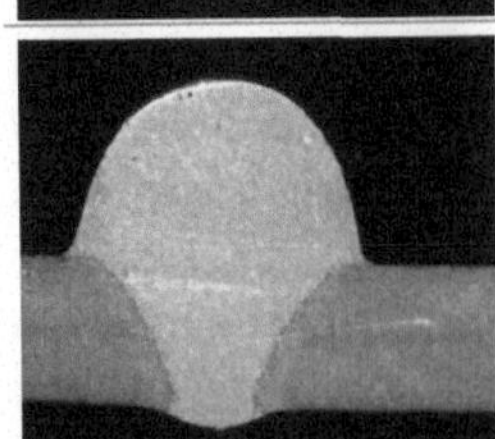	60	500	• Sufficient depth of penetration of filler metal • Moderate area of fusion zone • Weld bead contact angle $\leq 90°$ • Porosity density is moderate
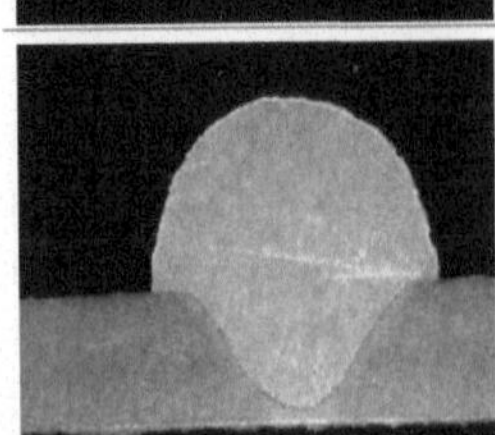	60	600	• Insufficient depth of penetration of filler metal • Moderate area of fusion zone • Weld bead contact angle $\leq 90°$ • Porosity density is moderate

(continued)

Table 2.8 (continued)

Macroscopic view of samples cross section	Iw (A)	Sw(mm/ min)	Remarks
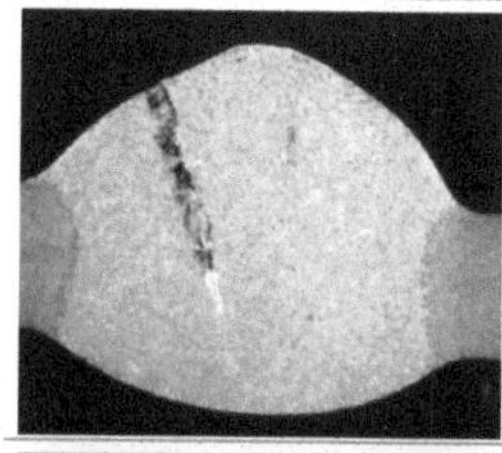	70	400	• Sufficient depth of penetration of filler metal • Larger area of fusion zone • Weld bead contact angle <90° • Porosity density is high
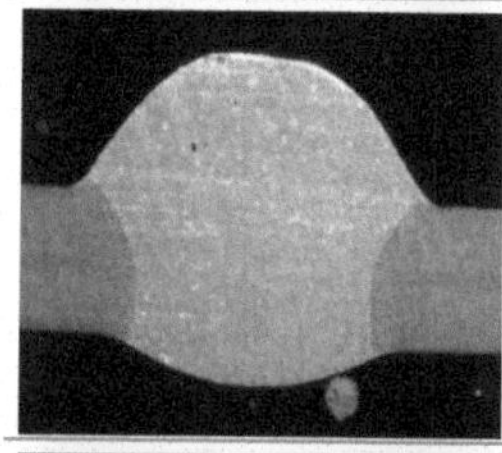	70	500	• Sufficient depth of penetration of filler metal • Large area of fusion zone • Weld bead contact angle <90° • Porosity density is high
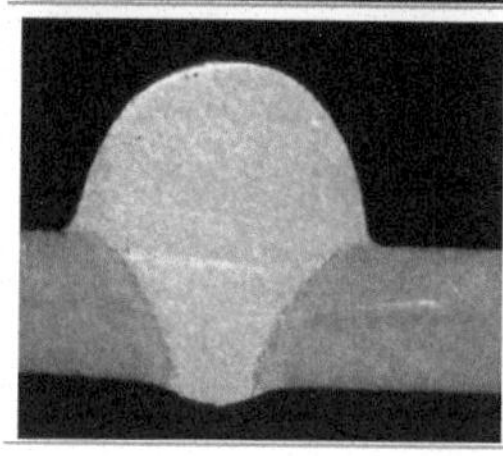	70	600	• Sufficient depth of penetration of filler metal • Moderate area of fusion zone • Weld bead contact angle <90° • Porosity density is high

$$Dilution\,(\%) = \frac{C^{M}_{\text{Coating}} - C^{M}_{\text{Filler}}}{C^{\text{Filler}}_{M}} \times 100 \tag{5.1}$$

where CM is the composition of metal M (M = Al, Si and Mn).

$$Dilution\,(\%) = \frac{zone1}{zone1 + zone2} \times 100 \tag{5.2}$$

$$\theta = Contact\,Angle$$

Dilution is not the most significant criteria in the study. However, it is imperative to estimate the quantum of base metal melts getting added to the bead. Bead angles are calculated as shown in Fig. 2.8, and an average of the two angles was reported. Table 2.9 shows the variation of the bead contact angle, the dilution and fusion zone area as a function of welding speed and welding current. With increase in welding speed, it is observed that the dilution reduces. A lower dilution is desire which will enhance the hardness of the weld. It is seen that as the speed ranges from 400 mm/min to 600 mm./min, the dilution drastically reduces from 62.8 to 6.7% for a constant current of 50 A. However, for higher currents, this difference is less

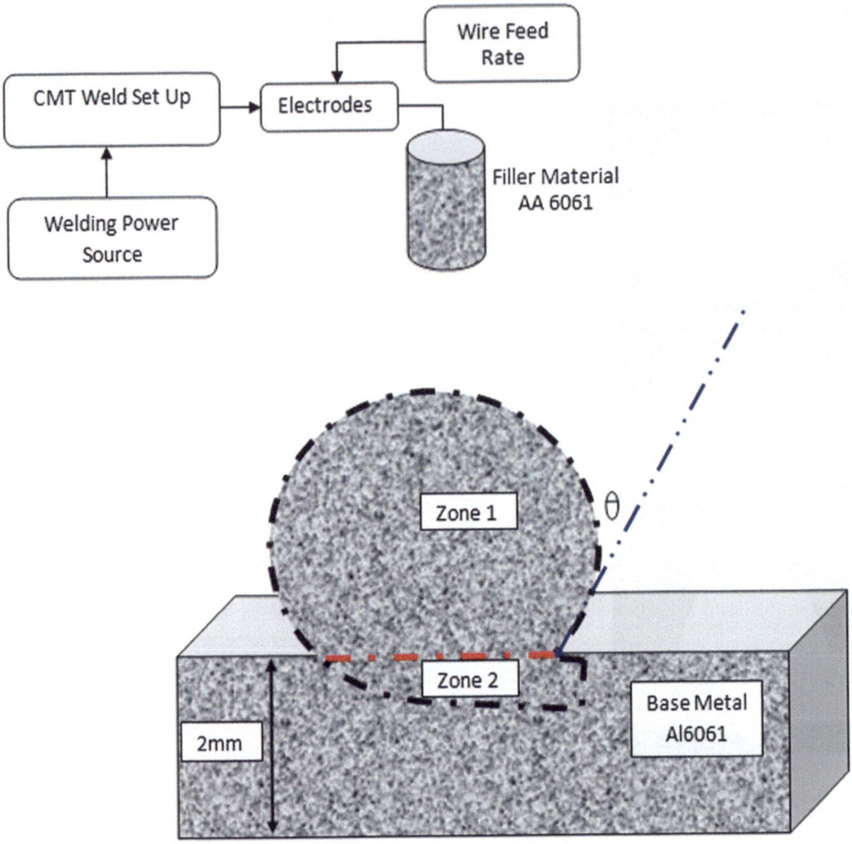

Fig. 2.8 Schematic showing how dilution was calculated and the bead angle definition [1]

compared with lower currents with increased speeds. The bead angle, however, increases from 38.7° to 111.5°. Thus, it is observed that the bead angle and dilution vary non-linearly with the welding speed. A bead angle of less than 90° is desirable so that there is no lack of fusion developed when the coating is made by side-by-side overlapping of beads. Thus, the welding speed of 500 mm/min is found to be the appropriate for overlay coatings in this case.

2.5.4.6 Effect of Welding Current and Welding Speed on Microstructures of CMT Welded AA6061

The microstructure characteristics of welded clads at the fusion line, fusion zone and heat-affected zone (HAZ) are essential in understanding the physical metallurgy of welding process and weld properties. The microstructures observed from the optical microscopy (OM) are shown in Fig. 2.9. The common precipitation sequence observed in Al–Mg–Si alloys is

$$SSSS \rightarrow GP \rightarrow \beta'' \rightarrow \beta' \rightarrow \beta - \mathrm{Mg_2Si}$$

Here SSSS represents the supersaturated solid solution and GP stands for Guinier–Preston zones. The supersaturated solid solution (SSSS) is a structure produced by solutionizing and quenching. During solutionizing, the alloy is first heated and then held at the temperature just below the eutectic temperature until the equilibrium solid solubility limit is attained. The purpose is to apply maximum amount of solutes (Mg and Si) responsible for strengthening into solid solution of the aluminium matrix. As a result of quenching (rapid cooling to room temperature), the solid solution becomes supersaturated. The resulting structure is soft and possesses the tendency to precipitate the solute available in abundance of the amount soluble at room temperature. This increases hardness and strength of the matrix, fine recrystallization of the aluminium grains occurs at the weld. Crystal grains of the AA6061 in the joint reduce in size considerably as compared to the size of the base metal. Nevertheless, the grains are coarsens and the density of grain boundaries is reduced in HAZ. Besides, in the weld, bending movements are induced as the crack propagates towards HAZ during fracture.

When the supersaturated solution is decomposed, the first products that are formed are Gunnier–Preston (G.P.) zones. Below 320 °F (160 °C) formation of G. P. zones are detected at temperatures under isothermal conditions. G.P. zones are referred to the formation as "preprecipitation" or "clustering," indicating true precipitation follows. Subsequently, material strengthening takes place when atoms of Mg and Si elements diffuse together. The diffusion of atoms results in either substitution or interstitial atomic defects in the crystal. The substitution of solute grains between the aluminium crystals results in hinder dislocation mobility of solute atoms, and hence the overall strength of alloy increases. The presence of Mg and Si clusters is clearly visible in the microstructures, with the increase in the current the heat-affected zone is slightly increased with broken clusters surrounding HAZ.

The precipitation of disperiods (Al_{12}–Mg_2 Cr) is because of presence of chromium as the grain refining agent during ingot pre-heat and high temperature homogenization treatments.

The chromium dispersoids assist holding the directional grain structure created/ generated during mechanical deformation (extrusion). The other advantage of chromium dispersoids is it limits the growth of recrystallized grains during the continuous heat treatments. Silicon as wetting agent and magnesium which improves the corrosion resistance both elements assist in balancing Al–Mg_2Si quasi-binary state. Mg and Si atoms diffuse together and form magnesium silicade β'(Mg_2Si) which is essential in improving the weldment strength. The hardening of these precipitates is taken place during artificial ageing of the alloy at temperature range of 433–463 K. The alloy is fully recrystallized with considerably large grains flattened in the extrusion or longitudinal direction. Clustering of the coarse

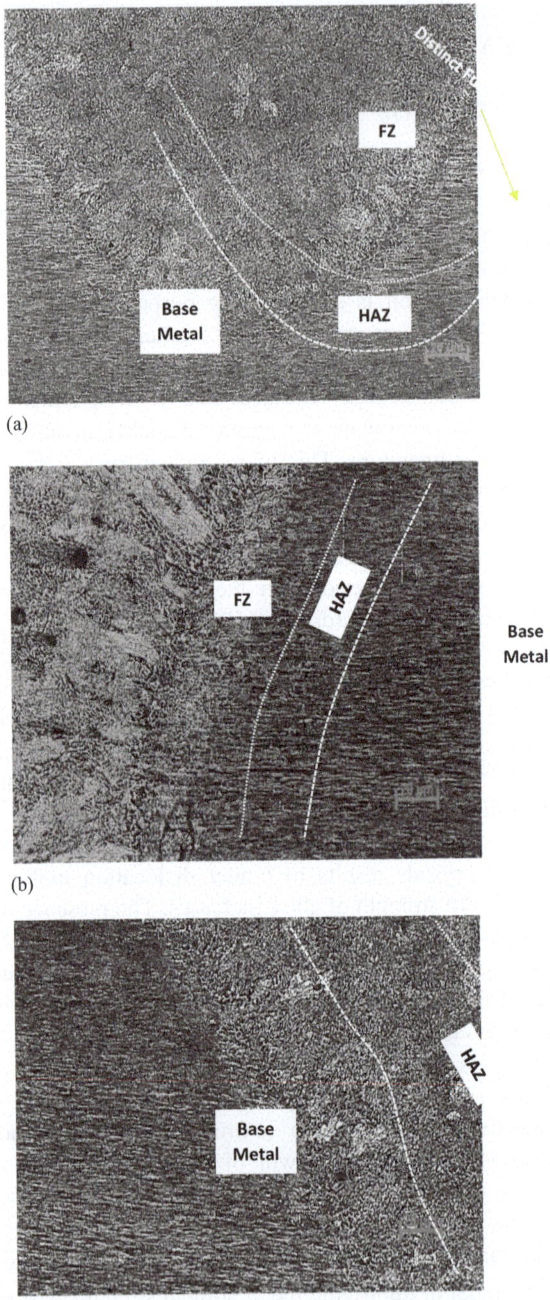

Fig. 2.9 Microstructures of Aluminium alloy 6061, as seen from Optical microscopy, clearly show fusion zone, fusion line and heat-affected zones (HAZ). **a** 50 A/400 mm/min, **b** 60 A/500 mm/min, **c** 70 A/500 mm/min [1]

constituents and other second-phase particles was observed. Cong et al. [13] observed the distribution of particles along the grain boundaries. Mg and Si being the key constituents of the Al alloy 6061, the base material indicates uniform distribution of silicates of Al with minute grain size. The characterization of the microstructure revealed epitaxial growth with dendrites aligned in columnar fusion emerging towards the source of heat.

2.6 Parametric Analysis Using Machine Learning Terminologies

This section presents details on CMT data analysis and predictions using machine learning techniques. The fundamental concepts based on which this work is performed have been discussed in Chap. 1 (Sect. 1.5).

2.6.1 Data Analysis Using NumPy *and* Pandas

Multiple datasets are used to predict Depth of Penetration, Reinforcement Height, Weld Pool Width, Heat input(J/mm) for CMT pulsed supply, Heat input(J/mm) for Continuous power supply, CMT_BW, CMT_DOP_2 and CMT_hi. The first step is to import the dataset. Figures 2.10, 2.11, 2.12, 2.13, 2.14, 2.15, 2.16 and 2.17 give an idea of how to import the dataset using Pandas (Fig. 2.18).

2.6.2 Data Visualization Using Seaborn *and* Matplotlib

The analysis and visualization done for the datasets are to determine the co-relation between the independent variables, i.e. inputs and dependent variable, i.e. output by the help of seaborn library. Figures 2.19, 2.20, 2.21, 2.22, 2.23, 2.24, 2.25 and 2.26 show the heat map and co-relation matrix for each dataset.

From the co-relation matrix of CMT_DOP mentioned above, the following input parameter welding current is positively co-related to Depth Of Penetration.

From the co-relation matrix of CMT_RH mentioned above, the following input parameter welding current negatively co-related to Reinforcement Height.

From the co-relation matrix of CMT_WPW mentioned above, the following input parameter welding current positively co-related to Weld pool width.

From the co-relation matrix of CMT_HI_CMTPS mentioned above, the following input parameters such as welding current and welding voltage are positively co-related to Heat input(J/mm) for CMT pulsed supply.

Fig. 2.10 Instructions to importing CMT_DOP

```
# Load the data and replace the '..' with nan
ts_df = pd.read_excel('CMT_DOP.xlsx')

ts_df = ts_df.astype(float)

display(ts_df)
```

	Welding Current (A)	Welding Speed (mm/min)	Depth of penetration (mm)
0	50.0	400.0	0.510000
1	50.0	500.0	0.330000
2	50.0	600.0	0.320000
3	60.0	400.0	2.680000
4	60.0	500.0	2.340000
5	60.0	600.0	1.670000
6	70.0	400.0	2.980000
7	70.0	500.0	2.560000
8	70.0	600.0	2.630000
9	80.0	400.0	4.375000
10	80.0	500.0	4.110000
11	80.0	600.0	3.858333
12	90.0	400.0	5.543333
13	90.0	500.0	5.285000
14	90.0	600.0	5.026667
15	100.0	400.0	6.711667
16	100.0	500.0	6.453333
17	100.0	600.0	6.195000
18	110.0	400.0	7.880000
19	110.0	500.0	7.621667
20	110.0	600.0	7.363333
21	120.0	400.0	9.048333
22	120.0	500.0	8.790000
23	120.0	600.0	8.531667
24	130.0	400.0	10.216667
25	130.0	500.0	9.958333
26	130.0	600.0	9.700000
27	140.0	400.0	11.385000
28	140.0	500.0	11.126667
29	140.0	600.0	10.868333
30	150.0	400.0	12.553333
31	150.0	500.0	12.295000
32	150.0	600.0	12.036667
33	160.0	400.0	13.721667
34	160.0	500.0	13.463333
35	160.0	600.0	13.205000
36	170.0	400.0	14.890067
37	170.0	500.0	14.631734
38	170.0	600.0	14.373401
39	180.0	400.0	16.058440
40	180.0	500.0	15.800106
41	180.0	600.0	15.541773

```
# Load the data and replace the '..' with nan
ts_df = pd.read_excel('CMT_RH.xlsx')

ts_df = ts_df.astype(float)

display(ts_df)
```

	Welding Current (A)	Welding Speed (mm/min)	Reinforcement height (mm)
0	50.0	400.0	2.690000
1	50.0	500.0	2.530000
2	50.0	600.0	2.510000
3	60.0	400.0	2.550000
4	60.0	500.0	2.500000
5	60.0	600.0	2.640000
6	70.0	400.0	2.140000
7	70.0	500.0	1.930000
8	70.0	600.0	2.060000
9	80.0	400.0	1.889444
10	80.0	500.0	1.861111
11	80.0	600.0	1.832778
12	90.0	400.0	1.622778
13	90.0	500.0	1.594444
14	90.0	600.0	1.566111
15	100.0	400.0	1.356111
16	100.0	500.0	1.327778
17	100.0	600.0	1.299444
18	110.0	400.0	1.089444
19	110.0	500.0	1.061111
20	110.0	600.0	1.032778
21	120.0	400.0	0.822778
22	120.0	500.0	0.794444
23	120.0	600.0	0.766111
24	130.0	400.0	0.556111
25	130.0	500.0	0.527778
26	130.0	600.0	0.499444
27	140.0	400.0	0.289444
28	140.0	500.0	0.261111
29	140.0	600.0	0.232778
30	150.0	400.0	0.022778
31	150.0	500.0	0.005556
32	150.0	600.0	0.033889
33	160.0	400.0	0.243889
34	160.0	500.0	0.272222
35	160.0	600.0	0.300556
36	170.0	400.0	0.510556
37	170.0	500.0	0.538889
38	170.0	600.0	0.567222
39	180.0	400.0	0.777222
40	180.0	500.0	0.805556
41	180.0	600.0	0.833889

Fig. 2.11 Instructions to import CMT_RH

Fig. 2.12 Instructions to import CMT_WPW

```
# Load the data and replace the '..' with nan
ts_df = pd.read_excel('CMT_WPW.xlsx')

ts_df = ts_df.astype(float)

display(ts_df)
```

	Welding Current (A)	Welding Speed (mm/min)	Weld pool width (mm)
0	50.0	400.0	3.220000
1	50.0	500.0	2.320000
2	50.0	600.0	1.750000
3	60.0	400.0	5.490000
4	60.0	500.0	4.890000
5	60.0	600.0	4.370000
6	70.0	400.0	6.800000
7	70.0	500.0	5.300000
8	70.0	600.0	5.610000
9	80.0	400.0	8.520000
10	80.0	500.0	7.890000
11	80.0	600.0	7.260000
12	90.0	400.0	10.256667
13	90.0	500.0	9.626667
14	90.0	600.0	8.996667
15	100.0	400.0	11.993333
16	100.0	500.0	11.363333
17	100.0	600.0	10.733333
18	110.0	400.0	13.730000
19	110.0	500.0	13.100000
20	110.0	600.0	12.470000
21	120.0	400.0	15.466667
22	120.0	500.0	14.836667
23	120.0	600.0	14.206667
24	130.0	400.0	17.203333
25	130.0	500.0	16.573333
26	130.0	600.0	15.943333
27	140.0	400.0	18.940000
28	140.0	500.0	18.310000
29	140.0	600.0	17.680000
30	150.0	400.0	20.676667
31	150.0	500.0	20.046667
32	150.0	600.0	19.416667
33	160.0	400.0	22.413333
34	160.0	500.0	21.783333
35	160.0	600.0	21.153333
36	170.0	400.0	24.150000
37	170.0	500.0	23.520000
38	170.0	600.0	22.890000
39	180.0	400.0	25.886667
40	180.0	500.0	25.256667
41	180.0	600.0	24.626667

```
# Load the data and replace the '..' with nan
ts_df = pd.read_excel('CMT_HI_CMTPS.xlsx')

ts_df = ts_df.astype(float)

display(ts_df)
```

	Welding current (A)	Welding voltage (V)	Welding speed (mm/min)	Heat input(J/mm) for CMT pulsed supply
0	50.0	11.7	400.0	67.625000
1	50.0	11.2	500.0	52.400000
2	50.0	11.0	600.0	43.250000
3	60.0	14.2	400.0	97.850000
4	60.0	13.2	500.0	73.250000
5	60.0	12.0	600.0	56.000000
6	70.0	13.2	400.0	79.960000
7	70.0	13.3	500.0	85.775000
8	70.0	13.7	600.0	55.940000
9	80.0	11.7	400.0	70.736494
10	80.0	11.2	500.0	54.764101
11	80.0	11.0	600.0	41.586388
12	90.0	14.2	400.0	93.977454
13	90.0	13.2	500.0	73.347258
14	90.0	12.0	600.0	50.853941
15	100.0	13.2	400.0	84.613799
16	100.0	13.3	500.0	74.230767
17	100.0	13.7	600.0	66.642417
18	110.0	11.7	400.0	70.592341
19	110.0	11.2	500.0	54.619947
20	110.0	11.0	600.0	41.442235
21	120.0	14.2	400.0	93.833300
22	120.0	13.2	500.0	73.203105
23	120.0	12.0	600.0	50.709788
24	130.0	13.2	400.0	84.469645
25	130.0	13.3	500.0	74.086614
26	130.0	13.7	600.0	66.498264
27	140.0	11.7	400.0	70.448188
28	140.0	11.2	500.0	54.475794
29	140.0	11.0	600.0	41.298081
30	150.0	14.2	400.0	93.689147
31	150.0	13.2	500.0	73.058951
32	150.0	12.0	600.0	50.565634
33	160.0	13.2	400.0	84.325492
34	160.0	13.3	500.0	73.942461
35	160.0	13.7	600.0	66.354111

Fig. 2.13 Instructions to import CMT_HI_CMTPS

```
# Load the data and replace the '..' with nan
ts_df = pd.read_excel('CMT_HI_CPS.xlsx')

ts_df = ts_df.astype(float)

display(ts_df)
```

	Welding current (A)	Welding voltage (V)	Welding speed (mm/min)	Heat input(J/mm) for Continuous power supply
0	50.0	11.7	400.0	87.700000
1	50.0	11.2	500.0	67.200000
2	50.0	11.0	600.0	55.000000
3	60.0	14.2	400.0	127.800000
4	60.0	13.2	500.0	95.000000
5	60.0	12.0	600.0	72.000000
6	70.0	13.2	400.0	138.600000
7	70.0	13.3	500.0	111.700000
8	70.0	13.7	600.0	95.900000
9	80.0	11.7	400.0	140.829574
10	80.0	11.2	500.0	118.346837
11	80.0	11.0	600.0	97.712312
12	90.0	14.2	400.0	172.479272
13	90.0	13.2	500.0	146.916179
14	90.0	12.0	600.0	120.120943
15	100.0	13.2	400.0	182.566481
16	100.0	13.3	500.0	163.780170
17	100.0	13.7	600.0	146.842072
18	110.0	11.7	400.0	189.573334
19	110.0	11.2	500.0	167.090597
20	110.0	11.0	600.0	146.456072
21	120.0	14.2	400.0	221.223032
22	120.0	13.2	500.0	195.659939
23	120.0	12.0	600.0	168.864703
24	130.0	13.2	400.0	231.310241
25	130.0	13.3	500.0	212.523930
26	130.0	13.7	600.0	195.585832
27	140.0	11.7	400.0	238.317094
28	140.0	11.2	500.0	215.834357
29	140.0	11.0	600.0	195.199832
30	150.0	14.2	400.0	269.966792
31	150.0	13.2	500.0	244.403699
32	150.0	12.0	600.0	217.608463
33	160.0	13.2	400.0	280.054001
34	160.0	13.3	500.0	261.267690
35	160.0	13.7	600.0	244.329592

Fig. 2.14 Instructions to importing CMT_HI_CPS

Fig. 2.15 Instructions to importing CMT_BW

```
# Load the data and replace the '..' with nan
ts_df = pd.read_excel('CMT_BW.xlsx')

ts_df = ts_df.astype(float)

display(ts_df)
```

	WC	WS	WV	BW
0	70.0	450.0	17.1	4.630000
1	70.0	500.0	17.1	4.350000
2	70.0	550.0	17.1	4.120000
3	90.0	450.0	18.3	5.560000
4	90.0	500.0	18.3	5.250000
5	90.0	550.0	18.3	4.910000
6	110.0	450.0	19.4	6.520000
7	110.0	500.0	19.4	6.150000
8	110.0	550.0	19.4	5.810000
9	130.0	450.0	17.1	8.978333
10	130.0	500.0	17.1	8.666667
11	130.0	550.0	17.1	8.355000
12	150.0	450.0	18.3	9.851667
13	150.0	500.0	18.3	9.540000
14	150.0	550.0	18.3	9.228333
15	170.0	450.0	19.4	10.771667
16	170.0	500.0	19.4	10.460000
17	170.0	550.0	19.4	10.148333
18	190.0	450.0	17.1	13.278333
19	190.0	500.0	17.1	12.966667
20	190.0	550.0	17.1	12.655000
21	210.0	450.0	18.3	14.151667
22	210.0	500.0	18.3	13.840000
23	210.0	550.0	18.3	13.528333
24	230.0	450.0	19.4	15.071667
25	230.0	500.0	19.4	14.760000
26	230.0	550.0	19.4	14.448333
27	250.0	450.0	17.1	17.578333
28	250.0	500.0	17.1	17.266667
29	250.0	550.0	17.1	16.955000
30	270.0	450.0	18.3	18.451667
31	270.0	500.0	18.3	18.140000
32	270.0	550.0	18.3	17.828333
33	290.0	450.0	19.4	19.371667
34	290.0	500.0	19.4	19.060000
35	290.0	550.0	19.4	18.748333

Dout

Fig. 2.16 Instructions to import CMT_DOP_2

```
# Load the data and replace the '..' with nan
ts_df = pd.read_excel('CMT_DOP_2.xlsx')

ts_df = ts_df.astype(float)

display(ts_df)
```

	WC	WS	WV	CMT_DOP_2
0	70.0	450.0	17.1	1.350000
1	70.0	500.0	17.1	0.850000
2	70.0	550.0	17.1	0.450000
3	90.0	450.0	18.3	3.110000
4	90.0	500.0	18.3	2.450000
5	90.0	550.0	18.3	1.650000
6	110.0	450.0	19.4	5.290000
7	110.0	500.0	19.4	4.230000
8	110.0	550.0	19.4	3.320000
9	130.0	450.0	17.1	19.005000
10	130.0	500.0	17.1	18.283333
11	130.0	550.0	17.1	17.561667
12	150.0	450.0	18.3	20.525000
13	150.0	500.0	18.3	19.803333
14	150.0	550.0	18.3	19.081667
15	170.0	450.0	19.4	22.401667
16	170.0	500.0	19.4	21.680000
17	170.0	550.0	19.4	20.958333
18	190.0	450.0	17.1	36.405000
19	190.0	500.0	17.1	35.683333
20	190.0	550.0	17.1	34.961667
21	210.0	450.0	18.3	37.925000
22	210.0	500.0	18.3	37.203333
23	210.0	550.0	18.3	36.481667
24	230.0	450.0	19.4	39.801667
25	230.0	500.0	19.4	39.080000
26	230.0	550.0	19.4	38.358333
27	250.0	450.0	17.1	53.805000
28	250.0	500.0	17.1	53.083333
29	250.0	550.0	17.1	52.361667
30	270.0	450.0	18.3	55.325000
31	270.0	500.0	18.3	54.603333
32	270.0	550.0	18.3	53.881667
33	290.0	450.0	19.4	57.201667
34	290.0	500.0	19.4	56.480000
35	290.0	550.0	19.4	55.758333

Fig. 2.17 Instructions to import CMT_HI

```
# Load the data and replace the '..' with nan
ts_df = pd.read_excel('CMT_HI.xlsx')

ts_df = ts_df.astype(float)

display(ts_df)
```

	WC	WS	WV	HI
0	70.0	450.0	17.1	0.143000
1	70.0	500.0	17.1	0.129000
2	70.0	550.0	17.1	0.117000
3	90.0	450.0	18.3	0.197000
4	90.0	500.0	18.3	0.177000
5	90.0	550.0	18.3	0.161000
6	110.0	450.0	19.4	0.256000
7	110.0	500.0	19.4	0.230000
8	110.0	550.0	19.4	0.209000
9	130.0	450.0	17.1	0.461833
10	130.0	500.0	17.1	0.443667
11	130.0	550.0	17.1	0.425500
12	150.0	450.0	18.3	0.510500
13	150.0	500.0	18.3	0.492333
14	150.0	550.0	18.3	0.474167
15	170.0	450.0	19.4	0.563833
16	170.0	500.0	19.4	0.545667
17	170.0	550.0	19.4	0.527500
18	190.0	450.0	17.1	0.775833
19	190.0	500.0	17.1	0.757667
20	190.0	550.0	17.1	0.739500
21	210.0	450.0	18.3	0.824500
22	210.0	500.0	18.3	0.806333
23	210.0	550.0	18.3	0.788167
24	230.0	450.0	19.4	0.877833
25	230.0	500.0	19.4	0.859667
26	230.0	550.0	19.4	0.841500
27	250.0	450.0	17.1	1.089833
28	250.0	500.0	17.1	1.071667
29	250.0	550.0	17.1	1.053500
30	270.0	450.0	18.3	1.138500
31	270.0	500.0	18.3	1.120333
32	270.0	550.0	18.3	1.102167
33	290.0	450.0	19.4	1.191833
34	290.0	500.0	19.4	1.173667
35	290.0	550.0	19.4	1.155500

```python
import seaborn as sns
import pandas as pd
import matplotlib.pyplot as plt
```

Fig. 2.18 Instructions to import libraries for data visualization and analysis

```python
# calculate the correlation matrix
corr = ts_df.corr()
# display the correlation matrix
display(corr)
# plot the correlation heatmap
sns.heatmap(corr, xticklabels=corr.columns, yticklabels=corr.columns, cmap='RdBu')
```

	Welding Current (A)	Welding Speed (mm/min)	Depth of penetration (mm)
Welding Current (A)	1.000000	0.000000	0.998377
Welding Speed (mm/min)	0.000000	1.000000	-0.044712
Depth of penetration (mm)	0.998377	-0.044712	1.000000

```
<matplotlib.axes._subplots.AxesSubplot at 0x15b928d1be0>
```

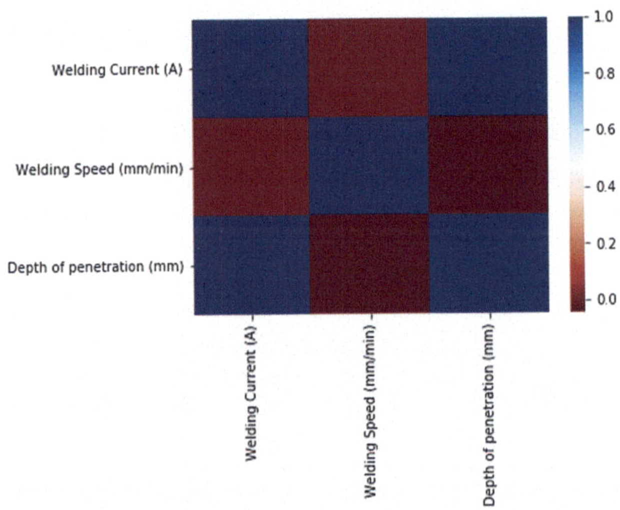

Fig. 2.19 CMT_DOP co-relation matrix

From the co-relation matrix of CMT_HI_CPS mentioned above, the following input parameters such as welding current and welding voltage are positively co-related to Heat input(J/mm) for Continuous power supply.

```
# calculate the correlation matrix
corr = ts_df.corr()
# display the correlation matrix
display(corr)
# plot the correlation heatmap
sns.heatmap(corr, xticklabels=corr.columns, yticklabels=corr.columns, cmap='RdBu')
```

	Welding Current (A)	Welding Speed (mm/min)	Reinforcement height (mm)
Welding Current (A)	1.00000	0.000000	-0.900530
Welding Speed (mm/min)	0.00000	1.000000	-0.013694
Reinforcement height (mm)	-0.90053	-0.013694	1.000000

```
<matplotlib.axes._subplots.AxesSubplot at 0x16758ef3ac8>
```

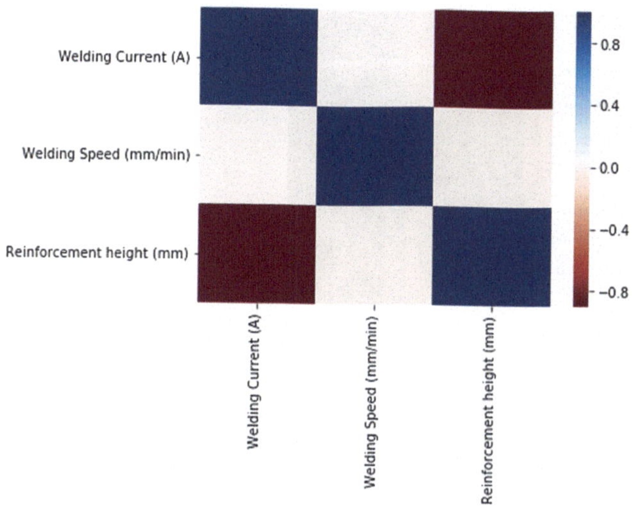

Fig. 2.20 CMT_RH co-relation matrix

From the co-relation matrix of CMT_HI_BW mentioned above, the following input parameters such as welding current and welding voltage are positively co-related to BW.

From the co-relation matrix of CMT_DOP_2 mentioned above, the following input parameter such as welding current is positively co-related to CMT_DOP_2.

From the co-relation matrix of CMT_HI mentioned above, the following input parameter such as welding current is positively co-related to CMT_HI.

```
# calculate the correlation matrix
corr = ts_df.corr()
# display the correlation matrix
display(corr)
# plot the correlation heatmap
sns.heatmap(corr, xticklabels=corr.columns, yticklabels=corr.columns, cmap='RdBu')
```

	Welding Current (A)	Welding Speed (mm/min)	Weld pool width (mm)
Welding Current (A)	1.000000	0.000000	0.996896
Welding Speed (mm/min)	0.000000	1.000000	-0.073249
Weld pool width (mm)	0.996896	-0.073249	1.000000

```
<matplotlib.axes._subplots.AxesSubplot at 0x2d15260d208>
```

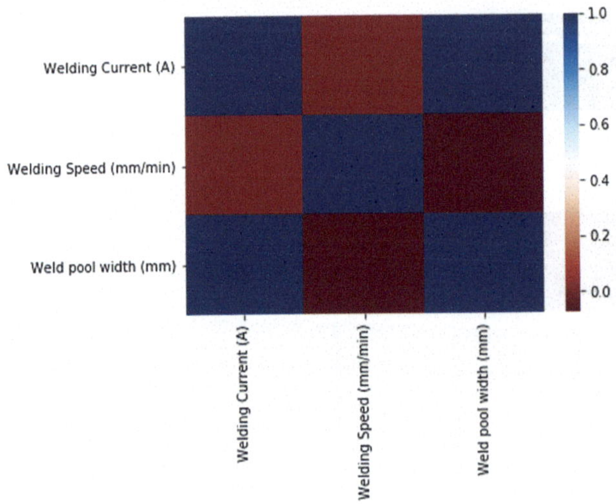

Fig. 2.21 CMT_WPW co-relation matrix

2.6.3 Data Preprocessing Using Scikit-Learn

Data preprocessing is one of the crucial steps in machine learning. This is an unavoidable step incase of data which are in unstructured form. Further to this, we will be discussing about data preprocessing by applying machine learning libraries such as scikit-learn.

This is a preprocessing technique used to derive non-linear features. It is mostly applied along with linear regression algorithm to train the model of higher degree. It is implemented in the following manner.

	Welding current (A)	Welding voltage (V)	Welding speed (mm/min)	Heat input(J/mm) for CMT pulsed supply
Welding current (A)	1.000000	0.186244	0.000000	0.108339
Welding voltage (V)	0.186244	1.000000	-0.299969	0.809993
Welding speed (mm/min)	0.000000	-0.299969	1.000000	-0.772135
Heat input(J/mm) for CMT pulsed supply	0.108339	0.809993	-0.772135	1.000000

```
<matplotlib.axes._subplots.AxesSubplot at 0x1e7e2aadd68>
```

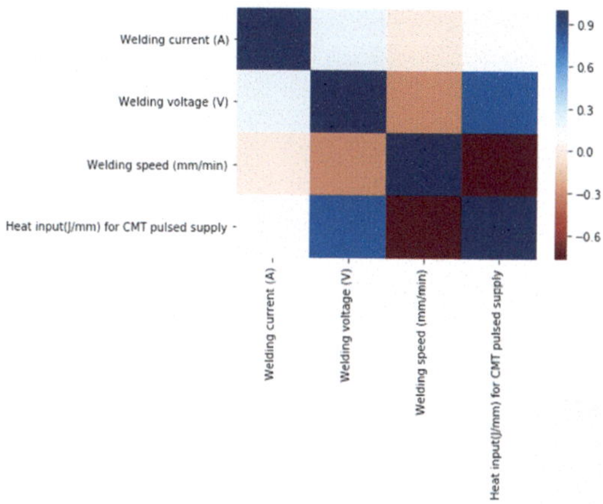

Fig. 2.22 CMT_HI_CMTPS co-relation matrix

```
# calculate the correlation matrix
corr = ts_df.corr()
# display the correlation matrix
display(corr)
# plot the correlation heatmap
sns.heatmap(corr, xticklabels=corr.columns, yticklabels=corr.columns, cmap='RdBu')
```

	Welding current (A)	Welding voltage (V)	Welding speed (mm/min)	Heat input(J/mm) for Continuous power supply
Welding current (A)	1.000000	0.186244	0.000000	0.949152
Welding voltage (V)	0.186244	1.000000	-0.299969	0.362623
Welding speed (mm/min)	0.000000	-0.299969	1.000000	-0.295550
Heat input(J/mm) for Continuous power supply	0.949152	0.362623	-0.295550	1.000000

```
<matplotlib.axes._subplots.AxesSubplot at 0x232b5844c50>
```

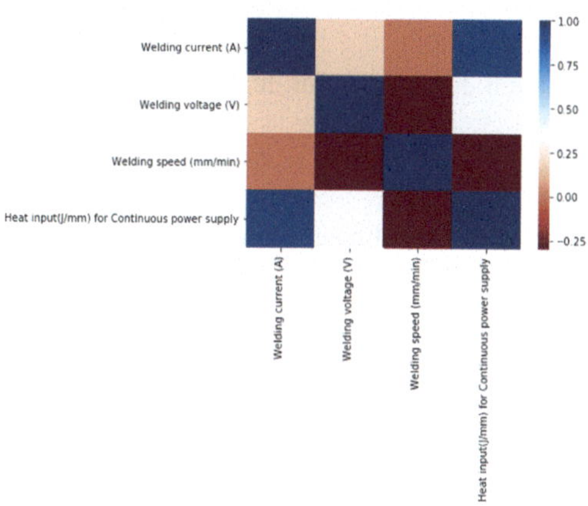

Fig. 2.23 CMT_HI_CPS co-relation matrix

```
# calculate the correlation matrix
corr = ts_df.corr()
# display the correlation matrix
display(corr)
# plot the correlation heatmap
sns.heatmap(corr, xticklabels=corr.columns, yticklabels=corr.columns, cmap='RdBu')
```

	WC	WS	WV	BW
WC	1.00000	0.000000	0.23645	0.994790
WS	0.00000	1.000000	0.00000	-0.052257
WV	0.23645	0.000000	1.00000	0.150240
BW	0.99479	-0.052257	0.15024	1.000000

```
<matplotlib.axes._subplots.AxesSubplot at 0x21f59d4a438>
```

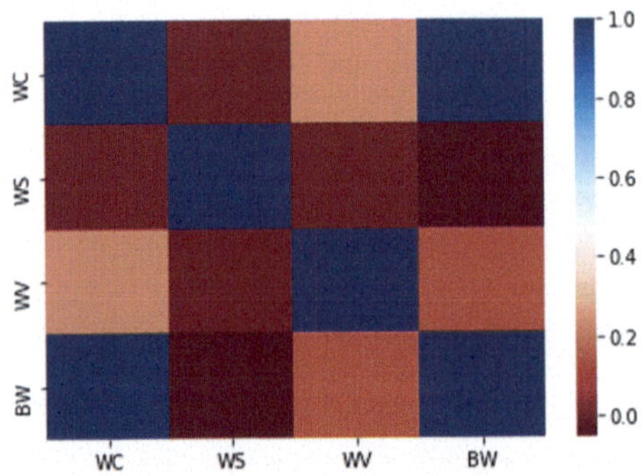

Fig. 2.24 CMT_BW co-relation matrix

Figure 2.27 shows the implementation for non-linear features of Depth of Penetration.

Figure 2.28 shows the implementation for non-linear features of Reinforcement Height.

Figure 2.29 shows the implementation for non-linear features of Weld pool width.

Figure 2.30 shows the implementation for non-linear features of Heat input for CMT pulsed supply.

Figure 2.31 shows the implementation for non-linear features of heat input for continuous power supply.

```
# calculate the correlation matrix
corr = ts_df.corr()
# display the correlation matrix
display(corr)
# plot the correlation heatmap
sns.heatmap(corr, xticklabels=corr.columns, yticklabels=corr.columns, cmap='RdBu')
```

	WC	WS	WV	CMT_DOP_2
WC	1.000000	0.000000	0.236450	0.985514
WS	0.000000	1.000000	0.000000	-0.030198
WV	0.236450	0.000000	1.000000	0.070936
CMT_DOP_2	0.985514	-0.030198	0.070936	1.000000

```
<matplotlib.axes._subplots.AxesSubplot at 0x1a21d83ee80>
```

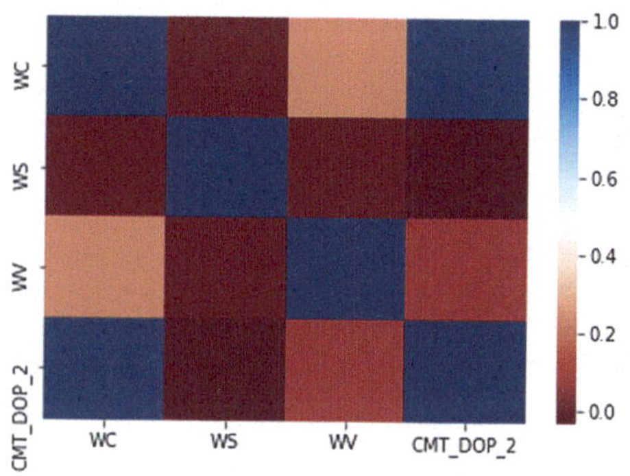

Fig. 2.25 CMT_DOP_2 co-relation matrix

Figure 2.32 shows the implementation for non-linear features of bead width.

Figure 2.33 shows the implementation for non-linear features of Depth of Penetration.

Figure 2.34 shows the implementation for non-linear features of Heat Index.

2.6.4 *Prediction Analysis Using Scikit-Learn*

In this section, we will look into various linear models for regression and classification using scikit-learn. The algorithms which will be discussed in the section to give a clarity about predictive analysis are the following.

```
# calculate the correlation matrix
corr = ts_df.corr()
# display the correlation matrix
display(corr)
# plot the correlation heatmap
sns.heatmap(corr, xticklabels=corr.columns, yticklabels=corr.columns, cmap='RdBu')
```

	WC	WS	WV	HI
WC	1.000000	0.00000	0.236450	0.991831
WS	0.000000	1.00000	0.000000	-0.041920
WV	0.236450	0.00000	1.000000	0.117569
HI	0.991831	-0.04192	0.117569	1.000000

<matplotlib.axes._subplots.AxesSubplot at 0x19997525e10>

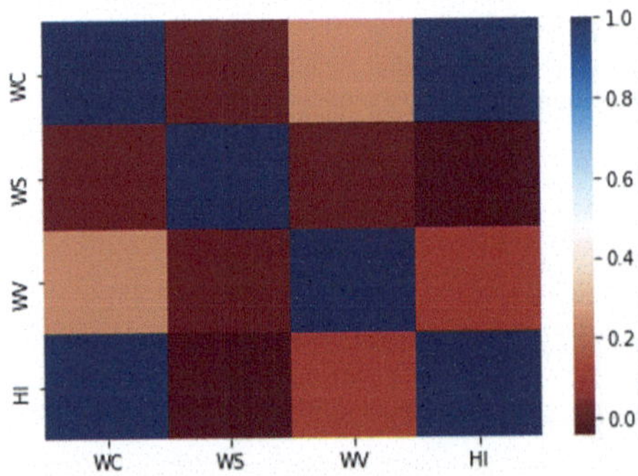

Fig. 2.26 CMT_HI co-relation matrix

Linear Regression

The fundamental concepts of linear regression along with the mathematical formulae based on which this work is performed have been discussed in Chap. 1 (Sect. 1.5).

A standard procedure of splitting the data into 80% of training data and 20% of testing data from the dataset is followed to predict the dependent parameters across each table. It is a two-step process, in which both the steps are a common process to split the dataset into training set and testing set (Fig. 2.35).

```
# get the summary
desc_df = ts_df.describe()
# add the standard deviation metric
desc_df.loc['+3_std'] = desc_df.loc['mean'] + (desc_df.loc['std'] * 3)
desc_df.loc['-3_std'] = desc_df.loc['mean'] - (desc_df.loc['std'] * 3)
# display it
desc_df
```

	Welding Current (A)	Welding Speed (mm/min)	Depth of penetration (mm)
count	42.000000	42.000000	42.000000
mean	115.000000	500.000000	8.205687
std	40.799928	82.639387	4.774700
min	50.000000	400.000000	0.320000
25%	80.000000	400.000000	4.176250
50%	115.000000	500.000000	8.205834
75%	150.000000	600.000000	12.230417
max	180.000000	600.000000	16.058440
+3_std	237.399785	747.918161	22.529787
-3_std	-7.399785	252.081839	-6.118413

```
# filter the data frame to remove the values exceeding 3 standard deviations
econ_remove_df = ts_df[(np.abs(stats.zscore(ts_df)) < 3).all(axis=1)]
# what rows were removed
ts_df.index.difference(econ_remove_df.index)
```

```
Int64Index([], dtype='int64')
```

Fig. 2.27 Instructions to preprocess the CMT_DOP Dataset

Figures 2.36, 2.37, 2.38, 2.39, 2.40, 2.41, 2.42 and 2.43 show the splitting of each dataset used for training the model.

Figure 2.36 shows that 80% of the data from the CMT_DOP Dataset will be used for training and the remaining 20% of the data will be used for testing the prediction of the trained model.

Figure 2.37 shows that 80% of the data from the CMT_RH Dataset will be used for training and the remaining 20% of the data will be used for testing the prediction of the trained model.

```
# get the summary
desc_df = ts_df.describe()
# add the standard deviation metric
desc_df.loc['+3_std'] = desc_df.loc['mean'] + (desc_df.loc['std'] * 3)
desc_df.loc['-3_std'] = desc_df.loc['mean'] - (desc_df.loc['std'] * 3)
# display it
desc_df
```

	Welding Current (A)	Welding Speed (mm/min)	Reinforcement height (mm)
count	42.000000	42.000000	42.000000
mean	115.000000	500.000000	1.160608
std	40.799928	82.639387	0.830945
min	50.000000	400.000000	0.005556
25%	80.000000	400.000000	0.514861
50%	115.000000	500.000000	0.933334
75%	150.000000	600.000000	1.854028
max	180.000000	600.000000	2.690000
+3_std	237.399785	747.918161	3.653443
-3_std	-7.399785	252.081839	-1.332226

```
# filter the data frame to remove the values exceeding 3 standard deviations
econ_remove_df = ts_df[(np.abs(stats.zscore(ts_df)) < 3).all(axis=1)]
# what rows were removed
ts_df.index.difference(econ_remove_df.index)
```

```
Int64Index([], dtype='int64')
```

Fig. 2.28 Instructions to preprocess the CMT_RH Dataset

Figure 2.38 shows that 80% of the data from the CMT_WPW Dataset will be used for training and the remaining 20% of the data will be used for testing the prediction of the trained model.

Figure 2.39 shows that 80% of the data from the CMT_HI_CMTPS Dataset will be used for training and the remaining 20% of the data will be used for testing the prediction of the trained model.

```
# get the summary
desc_df = ts_df.describe()
# add the standard deviation metric
desc_df.loc['+3_std'] = desc_df.loc['mean'] + (desc_df.loc['std'] * 3)
desc_df.loc['-3_std'] = desc_df.loc['mean'] - (desc_df.loc['std'] * 3)
# display it
desc_df
```

	Welding Current (A)	Welding Speed (mm/min)	Weld pool width (mm)
count	42.000000	42.000000	42.000000
mean	115.000000	500.000000	13.968333
std	40.799928	82.639387	7.107648
min	50.000000	400.000000	1.750000
25%	80.000000	400.000000	8.047500
50%	115.000000	500.000000	13.968333
75%	150.000000	600.000000	19.889167
max	180.000000	600.000000	25.886667
+3_std	237.399785	747.918161	35.291279
-3_std	-7.399785	252.081839	-7.354612

```
# filter the data frame to remove the values exceeding 3 standard deviations
econ_remove_df = ts_df[(np.abs(stats.zscore(ts_df)) < 3).all(axis=1)]
# what rows were removed
ts_df.index.difference(econ_remove_df.index)
```

```
Int64Index([], dtype='int64')
```

Fig. 2.29 Instructions to preprocess the CMT_WPW Dataset

Figure 2.40 shows that 80% of the data from the CMT_HI_CPS Dataset will be used for training and the remaining 20% of the data will be used for testing the prediction of the trained model.

Figure 2.41 shows that 80% of the data from the CMT_HI_BW Dataset will be used for training and the remaining 20% of the data will be used for testing the prediction of the trained model.

```
# get the summary
desc_df = ts_df.describe()
# add the standard deviation metric
desc_df.loc['+3_std'] = desc_df.loc['mean'] + (desc_df.loc['std'] * 3)
desc_df.loc['-3_std'] = desc_df.loc['mean'] - (desc_df.loc['std'] * 3)
# display it
desc_df
```

	Welding current (A)	Welding voltage (V)	Welding speed (mm/min)	Heat input(J/mm) for CMT pulsed supply
count	36.000000	36.000000	36.000000	36.000000
mean	105.000000	12.611111	500.000000	67.789325
std	35.010203	1.104220	82.807867	16.130580
min	50.000000	11.000000	400.000000	41.298081
25%	77.500000	11.700000	400.000000	54.583909
50%	105.000000	13.200000	500.000000	70.520264
75%	132.500000	13.300000	600.000000	75.663075
max	160.000000	14.200000	600.000000	97.850000
+3_std	210.030608	15.923771	748.423601	116.181065
-3_std	-0.030608	9.298451	251.576399	19.397586

```
# filter the data frame to remove the values exceeding 3 standard deviations
econ_remove_df = ts_df[(np.abs(stats.zscore(ts_df)) < 3).all(axis=1)]
# what rows were removed
ts_df.index.difference(econ_remove_df.index)
```

```
Int64Index([], dtype='int64')
```

Fig. 2.30 Instructions to preprocess the CMT_HI_CMTPS Dataset

```
# get the summary
desc_df = ts_df.describe()
# add the standard deviation metric
desc_df.loc['+3_std'] = desc_df.loc['mean'] + (desc_df.loc['std'] * 3)
desc_df.loc['-3_std'] = desc_df.loc['mean'] - (desc_df.loc['std'] * 3)
# display it
desc_df
```

	Welding current (A)	Welding voltage (V)	Welding speed (mm/min)	Heat input(J/mm) for Continuous power supply
count	36.000000	36.000000	36.000000	36.000000
mean	105.000000	12.611111	500.000000	167.660084
std	35.010203	1.104220	82.807867	61.266524
min	50.000000	11.000000	400.000000	55.000000
25%	77.500000	11.700000	400.000000	119.677416
50%	105.000000	13.200000	500.000000	167.977650
75%	132.500000	13.300000	600.000000	216.277884
max	160.000000	14.200000	600.000000	280.054001
+3_std	210.030608	15.923771	748.423601	351.459656
-3_std	-0.030608	9.298451	251.576399	-16.139487

```
# filter the data frame to remove the values exceeding 3 standard deviations
econ_remove_df = ts_df[(np.abs(stats.zscore(ts_df)) < 3).all(axis=1)]
# what rows were removed
ts_df.index.difference(econ_remove_df.index)
```

```
Int64Index([], dtype='int64')
```

Fig. 2.31 Instructions to preprocess the CMT_HI_CPS Dataset

```
# get the summary
desc_df = ts_df.describe()
# add the standard deviation metric
desc_df.loc['+3_std'] = desc_df.loc['mean'] + (desc_df.loc['std'] * 3)
desc_df.loc['-3_std'] = desc_df.loc['mean'] - (desc_df.loc['std'] * 3)
# display it
desc_df
```

	WC	WS	WV	BW
count	36.000000	36.000000	36.000000	36.000000
mean	180.000000	500.000000	18.266667	11.705556
std	70.020405	41.403934	0.952590	4.938748
min	70.000000	450.000000	17.100000	4.120000
25%	125.000000	450.000000	17.100000	7.896250
50%	180.000000	500.000000	18.300000	11.713334
75%	235.000000	550.000000	19.400000	15.542500
max	290.000000	550.000000	19.400000	19.371667
+3_std	390.061216	624.211801	21.124438	26.521800
-3_std	-30.061216	375.788199	15.408895	-3.110689

```
# filter the data frame to remove the values exceeding 3 standard deviations
econ_remove_df = ts_df[(np.abs(stats.zscore(ts_df)) < 3).all(axis=1)]
# what rows were removed
ts_df.index.difference(econ_remove_df.index)
```

```
Int64Index([], dtype='int64')
```

Fig. 2.32 Instructions to preprocess the CMT_BW Dataset

Figure 2.42 shows that 80% of the data from the CMT_DOP_2 Dataset will be used for training and the remaining 20% of the data will be used for testing the prediction of the trained model.

Figure 2.43 shows that 80% of the data from the CMT_HI Dataset will be used for training and the remaining 20% of the data will be used for testing the prediction of the trained model.

```
# get the summary
desc_df = ts_df.describe()
# add the standard deviation metric
desc_df.loc['+3_std'] = desc_df.loc['mean'] + (desc_df.loc['std'] * 3)
desc_df.loc['-3_std'] = desc_df.loc['mean'] - (desc_df.loc['std'] * 3)
# display it
desc_df
```

	WC	WS	WV	CMT_DOP_2
count	36.000000	36.000000	36.000000	36.000000
mean	180.000000	500.000000	18.266667	28.622222
std	70.020405	41.403934	0.952590	19.789227
min	70.000000	450.000000	17.100000	0.450000
25%	125.000000	450.000000	17.100000	14.493750
50%	180.000000	500.000000	18.300000	28.681667
75%	235.000000	550.000000	19.400000	42.941667
max	290.000000	550.000000	19.400000	57.201667
+3_std	390.061216	624.211801	21.124438	87.989904
-3_std	-30.061216	375.788199	15.408895	-30.745459

```
# filter the data frame to remove the values exceeding 3 standard deviations
econ_remove_df = ts_df[(np.abs(stats.zscore(ts_df)) < 3).all(axis=1)]
# what rows were removed
ts_df.index.difference(econ_remove_df.index)
```

```
Int64Index([], dtype='int64')
```

Fig. 2.33 Instructions to preprocess the CMT_DOP_2 Dataset

Once the model is trained, we use the trained model to predict the data in the following manner by the help of the following instructions (Fig. 2.44).

The predictions for each dataset are observed as shown in Tables 2.10, 2.11, 2.12, 2.13, 2.14, 2.15, 2.16 and 2.17.

From Fig. 2.45, it is observed that the actual value is quite in agreement with the predicted values of CMT_DOP.

From Fig. 2.46, it is observed that the actual value is quite in agreement with the predicted values of CMT_RH.

```
# get the summary
desc_df = ts_df.describe()
# add the standard deviation metric
desc_df.loc['+3_std'] = desc_df.loc['mean'] + (desc_df.loc['std'] * 3)
desc_df.loc['-3_std'] = desc_df.loc['mean'] - (desc_df.loc['std'] * 3)
# display it
desc_df
```

	WC	WS	WV	HI
count	36.000000	36.000000	36.000000	36.000000
mean	180.000000	500.000000	18.266667	0.650889
std	70.020405	41.403934	0.952590	0.358860
min	70.000000	450.000000	17.100000	0.117000
25%	125.000000	450.000000	17.100000	0.383125
50%	180.000000	500.000000	18.300000	0.651667
75%	235.000000	550.000000	19.400000	0.921750
max	290.000000	550.000000	19.400000	1.191833
+3_std	390.061216	624.211801	21.124438	1.727470
-3_std	-30.061216	375.788199	15.408895	-0.425692

```
# filter the data frame to remove the values exceeding 3 standard deviations
econ_remove_df = ts_df[(np.abs(stats.zscore(ts_df)) < 3).all(axis=1)]
# what rows were removed
ts_df.index.difference(econ_remove_df.index)
```

```
Int64Index([], dtype='int64')
```

Fig. 2.34 Instructions to preprocess the CMT_HI Dataset

```
# define our input variable (X) & output variable (Y)
X = ts_df.iloc[:, :-1]
Y = ts_df.iloc[:, 2]
Y

# Split X and y into X_
X_train, X_test, y_train, y_test = train_test_split(X, Y, test_size=.20, random_state=0)
#X_train, X_test, y_train, y_test = model_selection.train_test_split(X, y, train_size=0.75,test_size=0.25, random_state=101)
print ("X_train: ", X_train)
print ("y_train: ", y_train)
print("X_test: ", X_test)
print ("y_test: ", y_test)

# create a Linear Regression model object
regression_model = LinearRegression()

# pass through the X_train & y_train data set
regression_model.fit(X_train, y_train)
```

Fig. 2.35 Instructions to split the dataset into training and test data

```
# define our input variable (X) & output variable (Y)
X = ts_df.iloc[:, :-1]
Y = ts_df.iloc[:, 2]
Y

# Split X and y into X_
X_train, X_test, y_train, y_test = train_test_split(X, Y, test_size=.20, random_state=0)
#X_train, X_test, y_train, y_test = model_selection.train_test_split(X, y, train_size=0.75,test_size=0.25, random_state=101)
print ("X_train: ", X_train)
print ("y_train: ", y_train)
print("X_test: ", X_test)
print ("y_test: ", y_test)

# create a Linear Regression model object
regression_model = LinearRegression()

# pass through the X_train & y_train data set
regression_model.fit(X_train, y_train)
```

```
X_train:      Welding Current (A)  Welding Speed (mm/min)
31                  150.0                   500.0
29                  140.0                   600.0
20                  110.0                   600.0
39                  180.0                   400.0
2                    50.0                   600.0
41                  180.0                   600.0
18                  110.0                   400.0
15                  100.0                   400.0
22                  120.0                   500.0
16                  100.0                   500.0
38                  170.0                   600.0
8                    70.0                   600.0
13                   90.0                   500.0
5                    60.0                   600.0
17                  100.0                   600.0
32                  150.0                   600.0
14                   90.0                   600.0
35                  160.0                   600.0
7                    70.0                   500.0
34                  160.0                   500.0
1                    50.0                   500.0
26                  130.0                   600.0
12                   90.0                   400.0
33                  160.0                   400.0
24                  130.0                   400.0
6                    70.0                   400.0
23                  120.0                   600.0
21                  120.0                   400.0
19                  110.0                   500.0
9                    80.0                   400.0
40                  180.0                   500.0
3                    60.0                   400.0
0                    50.0                   400.0
y_train: 31    12.295000
29    10.868333
20     7.363333
39    16.058440
2      0.320000
41    15.541773
18     7.880000
15     6.711667
22     8.790000
16     6.453333
38    14.373401
8      2.630000
13     5.285000
5      1.670000
17     6.195000
32    12.036667
14     5.026667
35    13.205000
7      2.560000
34    13.463333
1      0.330000
26     9.700000
12     5.543333
33    13.721667
24    10.216667
6      2.980000
23     8.531667
21     9.048333
19     7.621667
9      4.375000
40    15.800106
3      2.680000
0      0.510000
Name: Depth of penetration (mm), dtype: float64
X_test:      Welding Current (A)  Welding Speed (mm/min)
30                 150.0                   400.0
36                 170.0                   400.0
27                 140.0                   400.0
4                   60.0                   500.0
10                  80.0                   500.0
25                 130.0                   500.0
28                 140.0                   500.0
11                  80.0                   600.0
37                 170.0                   500.0

y_test: 30    12.553333
36    14.890067
27    11.385000
4      2.340000
10     4.110000
25     9.958333
28    11.126667
11     3.858333
37    14.631734
Name: Depth of penetration (mm), dtype: float64

LinearRegression(copy_X=True, fit_intercept=True, n_jobs=None,
         normalize=False)
```

Fig. 2.36 Instructions to split CMT_DOP Dataset into 20% test and 80% training data

```
# Split X and y into X_
X_train, X_test, y_train, y_test = train_test_split(X, Y, test_size=.0, random_state=0)
#X_train, X_test, y_train, y_test = model_selection.train_test_split(X, y, train_size=0.75,test_size=0.25, random_state=101)
print ("X_train: ", X_train)
print ("y_train: ", y_train)
print("X_test: ", X_test)
print ("y_test: ", y_test)

# create a Linear Regression model object
regression_model = LinearRegression()

# pass through the X_train & y_train data set
regression_model.fit(X_train, y_train)
```

```
X_train:      Welding Current (A)  Welding Speed (mm/min)
30                   150                     400
36                   170                     400
27                   140                     400
4                     60                     500
10                    80                     500
25                   130                     500
28                   140                     500
11                    80                     600
37                   170                     500
31                   150                     500
29                   140                     600
20                   110                     600
39                   180                     400
2                     50                     600
41                   180                     600
18                   110                     400
15                   100                     400
22                   120                     500
16                   100                     500
38                   170                     600
8                     70                     600
13                    90                     500
5                     60                     600
17                   100                     600
32                   150                     600
14                    90                     600
35                   160                     600
7                     70                     500
34                   160                     500
1                     50                     500
26                   130                     600
12                    90                     400
33                   160                     400
24                   130                     400
6                     70                     400
23                   120                     600
21                   120                     400


19                   110                     500
9                     80                     400
40                   180                     500
3                     60                     400
0                     50                     400
y_train:  30    0.022778
36    0.510556
27    0.289444
4     2.500000
10    1.861111
25    0.527778
28    0.261111
11    1.832778
37    0.538889
31    0.005556
29    0.232778
20    1.032778
39    0.777222
2     2.510000
41    0.833889
18    1.089444
15    1.356111
22    0.794444
16    1.327778
38    0.567222
8     2.060000
13    1.594444
5     2.640000
17    1.299444
32    0.033889
14    1.566111
35    0.300556
7     1.930000
34    0.272222
1     2.530000
26    0.499444
12    1.622778
33    0.243889
24    0.556111
6     2.140000
23    0.766111
21    0.822778
19    1.061111
9     1.889444
40    0.805556
3     2.550000
0     2.690000
Name: Reinforcement height (mm), dtype: float64
X_test:  Empty DataFrame
Columns: [Welding Current (A), Welding Speed (mm/min)]
Index: []
y_test:  Series([], Name: Reinforcement height (mm), dtype: float64)

LinearRegression(copy_X=True, fit_intercept=True, n_jobs=None,
        normalize=False)
```

Fig. 2.37 Instructions to split CMT_RH Dataset into 20% test and 80% training data

```
# define our input variable (X) & output variable (Y)
X = ts_df.iloc[:, :-1]
Y = ts_df.iloc[:, 2]
Y

# Split X and y into X_
X_train, X_test, y_train, y_test = train_test_split(X, Y, test_size=.10, random_state=0)
#X_train, X_test, y_train, y_test = model_selection.train_test_split(X, y, train_size=0.75,test_size=0.25, random_state=101)
print ("X_train: ", X_train)
print ("y_train: ", y_train)
print("X_test: ", X_test)
print ("y_test: ", y_test)

# create a Linear Regression model object
regression_model = LinearRegression()

# pass through the X_train & y_train data set
regression_model.fit(X_train, y_train)
```

```
X_train:      Welding Current (A)  Welding Speed (mm/min)
25                  130.0                   500.0
28                  140.0                   500.0
11                   80.0                   600.0
37                  170.0                   500.0
31                  150.0                   500.0
29                  140.0                   600.0
20                  110.0                   600.0
39                  180.0                   400.0
2                    50.0                   600.0
41                  180.0                   600.0
18                  110.0                   400.0
15                  100.0                   400.0
22                  120.0                   500.0
16                  100.0                   500.0
38                  170.0                   600.0
8                    70.0                   600.0
13                   90.0                   500.0
5                    60.0                   600.0
17                  100.0                   600.0
32                  150.0                   600.0
14                   90.0                   600.0
35                  160.0                   600.0
7                    70.0                   500.0
34                  160.0                   500.0
1                    50.0                   500.0
26                  130.0                   600.0
12                   90.0                   400.0
33                  160.0                   400.0
24                  130.0                   400.0
6                    70.0                   400.0
23                  120.0                   600.0


21                  120.0                   400.0
19                  110.0                   500.0
9                    80.0                   400.0
40                  180.0                   500.0
3                    60.0                   400.0
0                    50.0                   400.0
y_train:  25    16.573333
28    18.310000
11     7.260000
37    23.520000
31    20.046667
29    17.680000
20    12.470000
39    25.886667
2      1.750000
41    24.626667
18    13.730000
15    11.993333
22    14.836667
16    11.363333
38    22.890000
8      5.610000
13     9.626667
5      4.370000
17    10.733333
32    19.416667
14     8.996667
35    21.153333
7      5.300000
34    21.783333
1      2.320000
26    15.943333
12    10.256667
33    22.413333
24    17.203333
6      6.800000
23    14.206667
21    15.466667
19    13.100000
9      8.520000
40    25.256667
3      5.490000
0      3.220000
Name: Weld pool width (mm), dtype: float64
X_test:      Welding Current (A)  Welding Speed (mm/min)
30                 150.0                   400.0
36                 170.0                   400.0
27                 140.0                   400.0
4                   60.0                   500.0
10                  80.0                   500.0
y_test:  30    20.676667
36    24.150000
27    18.940000
4      4.890000
10     7.890000
Name: Weld pool width (mm), dtype: float64

LinearRegression(copy_X=True, fit_intercept=True, n_jobs=None,
         normalize=False)
```

Fig. 2.38 Instructions to split CMT_WPW Dataset into 20% test and 80% training data

```
# define our input variable (X) & output variable (Y)
X = ts_df.iloc[:, :-1]
Y = ts_df.iloc[:, 3]
Y

# Split X and y into X_
X_train, X_test, y_train, y_test = train_test_split(X, Y, test_size=.20, random_state=0)
#X_train, X_test, y_train, y_test = model_selection.train_test_split(X, y, train_size=0.75,test_size=0.25, random_state=101)
print ("X_train: ", X_train)
print ("y_train: ", y_train)
print("X_test: ", X_test)
print ("y_test: ", y_test)

# create a Linear Regression model object
regression_model = LinearRegression()

# pass through the X_train & y_train data set
regression_model.fit(X_train, y_train)
```

```
X_train:      Welding current (A)  Welding voltage (V)  Welding speed (mm/min)
11                   80.0                 11.0                   600.0
29                  140.0                 11.0                   600.0
27                  140.0                 11.7                   400.0
35                  160.0                 13.7                   600.0
33                  160.0                 13.2                   400.0
28                  140.0                 11.2                   500.0
32                  150.0                 12.0                   600.0
8                    70.0                 13.7                   600.0
13                   90.0                 13.2                   500.0
5                    60.0                 12.0                   600.0
17                  100.0                 13.7                   600.0
14                   90.0                 12.0                   600.0
7                    70.0                 13.3                   500.0
26                  130.0                 13.7                   600.0
1                    50.0                 11.2                   500.0
12                   90.0                 14.2                   400.0
25                  130.0                 13.3                   500.0
24                  130.0                 13.2                   400.0
6                    70.0                 13.2                   400.0
23                  120.0                 12.0                   600.0
4                    60.0                 13.2                   500.0
18                  110.0                 11.7                   400.0
21                  120.0                 14.2                   400.0
19                  110.0                 11.2                   500.0
9                    80.0                 11.7                   400.0
34                  160.0                 13.3                   500.0
3                    60.0                 14.2                   400.0
0                    50.0                 11.7                   400.0
y_train:  11    41.586388

 29    41.298081
 27    70.448188
 35    66.354111
 33    84.325492
 28    54.475794
 32    50.565634
 8     55.940000
 13    73.347258
 5     56.000000
 17    66.642417
 14    50.853941
 7     85.775000
 26    66.498264
 1     52.400000
 12    93.977454
 25    74.086614
 24    84.469645
 6     79.960000
 23    50.709788
 4     73.250000
 18    70.592341
 21    93.833300
 19    54.619947
 9     70.736494
 34    73.942461
 3     97.850000
 0     67.625000
Name: Heat input(J/mm) for CMT pulsed supply, dtype: float64
X_test:      Welding current (A)  Welding voltage (V)  Welding speed (mm/min)
31                  150.0                 13.2                   500.0
20                  110.0                 11.0                   600.0
16                  100.0                 13.3                   500.0
30                  150.0                 14.2                   400.0
22                  120.0                 13.2                   500.0
15                  100.0                 13.2                   400.0
10                   80.0                 11.2                   500.0
2                    50.0                 11.0                   600.0
y_test:  31    73.058951
20    41.442235
16    74.230767
30    93.689147
22    73.203105
15    84.613799
10    54.764101
2     43.250000
Name: Heat input(J/mm) for CMT pulsed supply, dtype: float64

LinearRegression(copy_X=True, fit_intercept=True, n_jobs=None,
         normalize=False)
```

Fig. 2.39 Instructions to split CMT_HI_CMTPS Dataset into 20% test and 80% training data

```
# define our input variable (X) & output variable (Y)
X = ts_df.iloc[:, :-1]
Y = ts_df.iloc[:, 3]
Y

# Split X and y into X_
X_train, X_test, y_train, y_test = train_test_split(X, Y, test_size=.20, random_state=0)
#X_train, X_test, y_train, y_test = model_selection.train_test_split(X, y, train_size=0.75,test_size=0.25, random_state=101)
print ("X_train: ", X_train)
print ("y_train: ", y_train)
print("X_test: ", X_test)
print ("y_test: ", y_test)

# create a Linear Regression model object
regression_model = LinearRegression()

# pass through the X_train & y_train data set
regression_model.fit(X_train, y_train)
```

```
X_train:     Welding current (A)  Welding voltage (V)  Welding speed (mm/min)
11                 80.0                 11.0                  600.0
29                140.0                 11.0                  600.0
27                140.0                 11.7                  400.0
35                160.0                 13.7                  600.0
33                160.0                 13.2                  400.0
28                140.0                 11.2                  500.0
32                150.0                 12.0                  600.0
8                  70.0                 13.7                  600.0
13                 90.0                 13.2                  500.0
5                  60.0                 12.0                  600.0
17                100.0                 13.7                  600.0
14                 90.0                 12.0                  600.0
7                  70.0                 13.3                  500.0
26                130.0                 13.7                  600.0
1                  50.0                 11.2                  500.0
12                 90.0                 14.2                  400.0
25                130.0                 13.3                  500.0
24                130.0                 13.2                  400.0
6                  70.0                 13.2                  400.0
23                120.0                 12.0                  600.0
4                  60.0                 13.2                  500.0
18                110.0                 11.7                  400.0
21                120.0                 14.2                  400.0
19                110.0                 11.2                  500.0
9                  80.0                 11.7                  400.0
34                160.0                 13.3                  500.0
3                  60.0                 14.2                  400.0
0                  50.0                 11.7                  400.0
y_train:  11     97.712312

29    195.199832
27    238.317094
35    244.329592
33    280.054001
28    215.834357
32    217.608463
8      95.900000
13    146.916179
5      72.000000
17    146.842072
14    120.120943
7     111.700000
26    195.585832
1      67.200000
12    172.479272
25    212.523930
24    231.310241
6     138.600000
23    168.864703
4      95.000000
18    189.573334
21    221.223032
19    167.090597
9     140.829574
34    261.267690
3     127.800000
0      87.700000
Name: Heat input(J/mm) for Continuous power supply, dtype: float64
X_test:     Welding current (A)  Welding voltage (V)  Welding speed (mm/min)
31               150.0                 13.2                  500.0
20               110.0                 11.0                  600.0
16               100.0                 13.3                  500.0
30               150.0                 14.2                  400.0
22               120.0                 13.2                  500.0
15               100.0                 13.2                  400.0
10                80.0                 11.2                  500.0
2                 50.0                 11.0                  600.0
y_test:  31    244.403699
20    146.456072
16    163.780170
30    269.966792
22    195.659939
15    182.566481
10    118.346837
2      55.000000
Name: Heat input(J/mm) for Continuous power supply, dtype: float64

LinearRegression(copy_X=True, fit_intercept=True, n_jobs=None,
         normalize=False)
```

Fig. 2.40 Instructions to split CMT_HI_CPS Dataset into 20% test and 80% training data

```
# define our input variable (X) & output variable (Y)
X = ts_df.iloc[:, :-1]
Y = ts_df.iloc[:, 3]
Y

# Split X and y into X_
X_train, X_test, y_train, y_test = train_test_split(X, Y, test_size=.20, random_state=0)
#X_train, X_test, y_train, y_test = model_selection.train_test_split(X, y, train_size=0.75,test_size=0.25, random_state=101)
print ("X_train: ", X_train)
print ("y_train: ", y_train)
print("X_test: ", X_test)
print ("y_test: ", y_test)

# create a Linear Regression model object
regression_model = LinearRegression()

# pass through the X_train & y_train data set
regression_model.fit(X_train, y_train)
```

```
X_train:          WC     WS    WV
11  130.0  550.0  17.1
29  250.0  550.0  17.1
27  250.0  450.0  17.1
35  290.0  550.0  19.4
33  290.0  450.0  19.4
28  250.0  500.0  17.1
32  270.0  550.0  18.3
8   110.0  550.0  19.4
13  150.0  500.0  18.3
5    90.0  550.0  18.3
17  170.0  550.0  19.4
14  150.0  550.0  18.3
7   110.0  500.0  19.4
26  230.0  550.0  19.4
1    70.0  500.0  17.1
12  150.0  450.0  18.3
25  230.0  500.0  19.4
24  230.0  450.0  19.4
6   110.0  450.0  19.4
23  210.0  550.0  18.3
4    90.0  500.0  18.3
18  190.0  450.0  17.1
21  210.0  450.0  18.3
19  190.0  500.0  17.1
9   130.0  450.0  17.1
34  290.0  500.0  19.4
3    90.0  450.0  18.3
0    70.0  450.0  17.1

y_train:  11     8.355000
29     16.955000
27     17.578333
35     18.748333
33     19.371667
28     17.266667
32     17.828333
8       5.810000
13      9.540000
5       4.910000
17     10.148333
14      9.228333
7       6.150000
26     14.448333
1       4.350000
12      9.851667
25     14.760000
24     15.071667
6       6.520000
23     13.528333
4       5.250000
18     13.278333
21     14.151667
19     12.966667
9       8.978333
34     19.060000
3       5.560000
0       4.630000
Name: BW, dtype: float64
X_test:          WC     WS    WV
31  270.0  500.0  18.3
20  190.0  550.0  17.1
16  170.0  500.0  19.4
30  270.0  450.0  18.3
22  210.0  500.0  18.3
15  170.0  450.0  19.4
10  130.0  500.0  17.1
2    70.0  550.0  17.1
y_test:  31    18.140000
20     12.655000
16     10.460000
30     18.451667
22     13.840000
15     10.771667
10      8.666667
2       4.120000
Name: BW, dtype: float64

LinearRegression(copy_X=True, fit_intercept=True, n_jobs=None,
         normalize=False)
```

Fig. 2.41 Instructions to split CMT_HI_CPS Dataset into 20% test and 80% training data

```
# define our input variable (X) & output variable (Y)
X = ts_df.iloc[:, :-1]
Y = ts_df.iloc[:, 3]
Y

# Split X and y into X_
X_train, X_test, y_train, y_test = train_test_split(X, Y, test_size=.20, random_state=0)
#X_train, X_test, y_train, y_test = model_selection.train_test_split(X, y, train_size=0.75,test_size=0.25, random_state=101)
print ("X_train: ", X_train)
print ("y_train: ", y_train)
print("X_test: ", X_test)
print ("y_test: ", y_test)

# create a Linear Regression model object
regression_model = LinearRegression()

# pass through the X_train & y_train data set
regression_model.fit(X_train, y_train)
```

```
X_train:        WC     WS    WV
11  130.0  550.0  17.1
29  250.0  550.0  17.1
27  250.0  450.0  17.1
35  290.0  550.0  19.4
33  290.0  450.0  19.4
28  250.0  500.0  17.1
32  270.0  550.0  18.3
8   110.0  550.0  19.4
13  150.0  500.0  18.3
5    90.0  550.0  18.3
17  170.0  550.0  19.4
14  150.0  550.0  18.3
7   110.0  500.0  19.4
26  230.0  550.0  19.4
1    70.0  500.0  17.1
12  150.0  450.0  18.3
25  230.0  500.0  19.4
24  230.0  450.0  19.4
6   110.0  450.0  19.4
23  210.0  550.0  18.3
4    90.0  500.0  18.3
18  190.0  450.0  17.1
21  210.0  450.0  18.3
19  190.0  500.0  17.1
9   130.0  450.0  17.1
34  290.0  500.0  19.4
3    90.0  450.0  18.3
0    70.0  450.0  17.1

y_train: 11    17.561667
29    52.361667
27    53.805000
35    55.758333
33    57.201667
28    53.083333
32    53.881667
8      3.320000
13    19.803333
5      1.650000
17    20.958333
14    19.081667
7      4.230000
26    38.358333
1      0.850000
12    20.525000
25    39.080000
24    39.801667
6      5.290000
23    36.481667
4      2.450000
18    36.405000
21    37.925000
19    35.683333
9     19.005000
34    56.480000
3      3.110000
0      1.350000
Name: CMT_DOP_2, dtype: float64
X_test:        WC     WS    WV
31  270.0  500.0  18.3
20  190.0  550.0  17.1
16  170.0  500.0  19.4
30  270.0  450.0  18.3
22  210.0  500.0  18.3
15  170.0  450.0  19.4
10  130.0  500.0  17.1
2    70.0  550.0  17.1
y_test: 31    54.603333
20    34.961667
16    21.680000
30    55.325000
22    37.203333
15    22.401667
10    18.283333
2      0.450000
Name: CMT_DOP_2, dtype: float64

LinearRegression(copy_X=True, fit_intercept=True, n_jobs=None,
         normalize=False)
```

Fig. 2.42 Instructions to split CMT_DOP_2 Dataset into 20% test and 80% training data

```
# define our input variable (X) & output variable (Y)
X = ts_df.iloc[:, :-1]
Y = ts_df.iloc[:, 3]
Y

# Split X and y into X_
X_train, X_test, y_train, y_test = train_test_split(X, Y, test_size=.20, random_state=0)
#X_train, X_test, y_train, y_test = model_selection.train_test_split(X, y, train_size=0.75,test_size=0.25, random_state=101)
print ("X_train: ", X_train)
print ("y_train: ", y_train)
print("X_test: ", X_test)
print ("y_test: ", y_test)

# create a Linear Regression model object
regression_model = LinearRegression()

# pass through the X_train & y_train data set
regression_model.fit(X_train, y_train)
```

```
X_train:        WC     WS    WV
11  130.0  550.0  17.1
29  250.0  550.0  17.1
27  250.0  450.0  17.1
35  290.0  550.0  19.4
33  290.0  450.0  19.4
28  250.0  500.0  17.1
32  270.0  550.0  18.3
8   110.0  550.0  19.4
13  150.0  500.0  18.3
5    90.0  550.0  18.3
17  170.0  550.0  19.4
14  150.0  550.0  18.3
7   110.0  500.0  19.4
26  230.0  550.0  19.4
1    70.0  500.0  17.1
12  150.0  450.0  18.3
25  230.0  500.0  19.4
24  230.0  450.0  19.4
6   110.0  450.0  19.4
23  210.0  550.0  18.3
4    90.0  500.0  18.3
18  190.0  450.0  17.1
21  210.0  450.0  18.3
19  190.0  500.0  17.1
9   130.0  450.0  17.1
34  290.0  500.0  19.4
3    90.0  450.0  18.3
0    70.0  450.0  17.1

y_train:  11    0.425500
29    1.053500
27    1.089833
35    1.155500
33    1.191833
28    1.071667
32    1.102167
8     0.209000
13    0.492333
5     0.161000
17    0.527500
14    0.474167
7     0.230000
26    0.841500
1     0.129000
12    0.510500
25    0.859667
24    0.877833
6     0.256000
23    0.788167
4     0.177000
18    0.775833
21    0.824500
19    0.757667
9     0.461833
34    1.173667
3     0.197000
0     0.143000
Name: HI, dtype: float64
X_test:        WC     WS    WV
31  270.0  500.0  18.3
20  190.0  550.0  17.1
16  170.0  500.0  19.4
30  270.0  450.0  18.3
22  210.0  500.0  18.3
15  170.0  450.0  19.4
10  130.0  500.0  17.1
2    70.0  550.0  17.1
y_test:  31    1.120333
20    0.739500
16    0.545667
30    1.138500
22    0.806333
15    0.563833
10    0.443667
2     0.117000
Name: HI, dtype: float64

LinearRegression(copy_X=True, fit_intercept=True, n_jobs=None,
         normalize=False)
```

Fig. 2.43 Instructions to split CMT_HI Dataset into 20% test and 80% training data

Table 2.9 Measurements of weld beads

S. no.	Welding current (A)	Welding speed (mm/s)	Fusion zone area (mm^2)	Dilution based on area (%)	Bead contact angle (°)
A	50	400	8.739	6.67	111.5
B	60	400	7.670	5.6	112.6
C	70	400	6.040	4.9	118.8
D	50	500	17.941	42.9	59.7
E	60	500	15.947	34.8	61
F	70	500	162	27.6	61.3
G	50	600	23.413	62.8	38.7
H	60	600	19.734	60.8	37.5
I	70	600	17.930	58.3	40

```
# Get multiple predictions
y_predict = regression_model.predict(X)

# Show the predicions
y_predict[:]

df = pd.DataFrame({'Actual': Y, 'Predicted': y_predict})
df
```

Fig. 2.44 Instructions to predict by using the trained model

From Fig. 2.47, it is observed that the actual value is quite in agreement with the predicted values of CMT_WPW.

From Fig. 2.48, it is observed that the actual value is quite in agreement with the predicted values of CMT_HI_CMTPS.

From Fig. 2.49, it is observed that the actual value is quite in agreement with the predicted values of CMT_HI_CPS.

From Fig. 2.50, it is observed that the actual value is quite in agreement with the predicted values of CMT_BW.

Table 2.10 Results of CMT_DOP_predicted data

CMT_DOP_2	CMT_DOP_Predicted
1.35	1.574503
0.85	0.834215
0.45	0.093927
3.11	3.114528
2.45	2.37424
1.65	1.633952
5.29	5.009903
4.23	4.269615
3.32	3.529326
19.005	18.987172
18.283333	18.246884
17.561667	17.506596
20.525	20.527197
19.803333	19.786909
19.081667	19.046621
22.401667	22.422572
21.68	21.682284
20.958333	20.941996
36.405	36.399841
35.683333	35.659553
34.961667	34.919265
37.925	37.939866
37.203333	37.199578
36.481667	36.45929
39.801667	39.835241
39.08	39.094953
38.358333	38.354665
53.805	53.81251
53.083333	53.072222
52.361667	52.331934
55.325	55.352535
54.603333	54.612247
53.881667	53.871959
57.201667	57.24791
56.48	56.507622
55.758333	55.767334

Table 2.11 Results of RH_predicted data

RH	RH_Predicted
2.69	2.36651
2.53	2.352741
2.51	2.338971
2.55	2.183106
2.5	2.169336
2.64	2.155566
2.14	1.999701
1.93	1.985931
2.06	1.972161
1.889444	1.816296
1.861111	1.802526
1.832778	1.788756
1.622778	1.632891
1.594444	1.619121
1.566111	1.605351
1.356111	1.449486
1.327778	1.435716
1.299444	1.421946
1.089444	1.266081
1.061111	1.252311
1.032778	1.238541
0.822778	1.082676
0.794444	1.068906
0.766111	1.055136
0.556111	0.899271
0.527778	0.885501
0.499444	0.871731
0.289444	0.715866
0.261111	0.702096
0.232778	0.688326
0.022778	0.532461
0.005556	0.518691
0.033889	0.504921
0.243889	0.349056
0.272222	0.321516
0.300556	0.165651
0.510556	0.151881
0.538889	0.138111
0.567222	−0.017754
0.777222	−0.031524
0.805556	−0.045294
0.833889	−0.045294

Table 2.12 Results of WPW_predicted data

Weld pool width (mm)	WPW_Predicted
3.22	3.269712
2.32	2.639201
1.75	2.00869
5.49	5.010739
4.89	4.380228
4.37	3.749717
6.8	6.751766
5.3	6.121255
5.61	5.490744
8.52	8.492793
7.89	7.862282
7.26	7.231771
10.26	10.23
9.63	9.60
9.00	8.97
11.99	11.97
11.36	11.34
10.73	10.71
13.73	13.72
13.10	13.09
12.47	12.45
15.47	15.46
14.84	14.83
14.21	14.20
17.20	17.20
16.57	16.57
15.94	15.94
18.94	18.94
18.31	18.31
17.68	17.68
20.68	20.68
20.05	20.05
19.42	19.42
22.41	22.42
21.78	21.79
21.15	21.16
24.15	24.16
23.52	23.53
22.89	22.90
25.89	25.90
25.26	25.27
24.63	24.64

Table 2.13 Results of CMT_HI_CMTPS_P predicted data

Heat input (J/mm) for CMT pulsed supply	CMT_HI_CMTPS_P
67.625	70.702997
52.4	54.62424
43.25	41.360036
97.85	94.134395
73.25	73.364718
56	50.718673
79.96	84.729351
85.775	74.279699
55.94	66.644598
70.736494	70.633387
54.764101	54.55463
41.586388	41.290425
93.977454	94.064785
73.347258	73.295108
50.853941	50.649063
84.613799	84.659741
74.230767	74.210088
66.642417	66.574988
70.592341	70.563776
54.619947	54.48502
41.442235	41.220815
93.8333	93.995174
73.203105	73.225497
50.709788	50.579452
84.469645	84.59013
74.086614	74.140478
66.498264	66.505378
70.448188	70.494166
54.475794	54.415409
41.298081	41.151205
93.689147	93.925564
73.058951	73.155887
50.565634	50.509842
84.325492	84.52052
73.942461	74.070868
66.354111	66.435767

Table 2.14 Results of CMT_HI_CPS_P predicted data

Heat input (J/mm) for continuous power supply	CMT_HI_CPS_Predicted
87.7	91.380635
67.2	68.475692
55	47.49784
127.8	123.786272
95	97.669513
72	70.268027
138.6	133.70919
111.7	114.658428
95.9	97.534757
140.829574	140.420291
118.346837	117.515349
97.712312	96.537497
172.479272	172.825929
146.916179	146.709169
120.120943	119.307683
182.566481	182.748847
163.78017	163.698085
146.842072	146.574413
189.573334	189.459948
167.090597	166.555006
146.456072	145.577154
221.223032	221.865585
195.659939	195.748826
168.864703	168.34734
231.310241	231.788504
212.52393	212.737742
195.585832	195.61407
238.317094	238.499605
215.834357	215.594662
195.199832	194.61681
269.966792	270.905242
244.403699	244.788483
217.608463	217.386997
280.054001	280.82816
261.26769	261.777398
244.329592	244.653727

Table 2.15 Results of CMT_BW_P predicted data

BW	BW_P
4.63	4.671458
4.35	4.355594
4.12	4.039729
5.56	5.549306
5.25	5.233441
4.91	4.917577
6.52	6.473523
6.15	6.157659
5.81	5.841794
8.978333	8.974314
8.666667	8.65845
8.355	8.342585
9.851667	9.852162
9.54	9.536297
9.228333	9.220433
10.771667	10.77638
10.46	10.460515
10.148333	10.14465
13.278333	13.27717
12.966667	12.961306
12.655	12.645441
14.151667	14.155018
13.84	13.839153
13.528333	13.523289
15.071667	15.079236
14.76	14.763371
14.448333	14.447506
17.578333	17.580026
17.266667	17.264162
16.955	16.948297
18.451667	18.457874
18.14	18.142009
17.828333	17.826145
19.371667	19.382092
19.06	19.066227
18.748333	18.750362

Table 2.16 Results of CMT_DOP_2_predicted data

CMT_DOP_2	CMT_DOP_2_Predicted
1.35	1.574503
0.85	0.834215
0.45	0.093927
3.11	3.114528
2.45	2.37424
1.65	1.633952
5.29	5.009903
4.23	4.269615
3.32	3.529326
19.005	18.987172
18.283333	18.246884
17.561667	17.506596
20.525	20.527197
19.803333	19.786909
19.081667	19.046621
22.401667	22.422572
21.68	21.682284
20.958333	20.941996
36.405	36.399841
35.683333	35.659553
34.961667	34.919265
37.925	37.939866
37.203333	37.199578
36.481667	36.45929
39.801667	39.835241
39.08	39.094953
38.358333	38.354665
53.805	53.81251
53.083333	53.072222
52.361667	52.331934
55.325	55.352535
54.603333	54.612247
53.881667	53.871959
57.201667	57.24791
56.48	56.507622
55.758333	55.767334

Table 2.17 Results of CMT_HI_predicted data

HI	HI_P
0.143	0.147251
0.129	0.12873
0.117	0.110208
0.197	0.1963
0.177	0.177778
0.161	0.159257
0.256	0.24999
0.23	0.231469
0.209	0.212947
0.461833	0.461493
0.443667	0.442971
0.4255	0.42445
0.5105	0.510542
0.492333	0.49202
0.474167	0.473498
0.563833	0.564232
0.545667	0.54571
0.5275	0.527188
0.775833	0.775735
0.757667	0.757213
0.7395	0.738691
0.8245	0.824783
0.806333	0.806262
0.788167	0.78774
0.877833	0.878474
0.859667	0.859952
0.8415	0.84143
1.089833	1.089976
1.071667	1.071455
1.0535	1.052933
1.1385	1.139025
1.120333	1.120503
1.102167	1.101982
1.191833	1.192715
1.173667	1.174194
1.1555	1.155672

Fig. 2.45 CMT_DOP V/S CMT_DOP_predicted

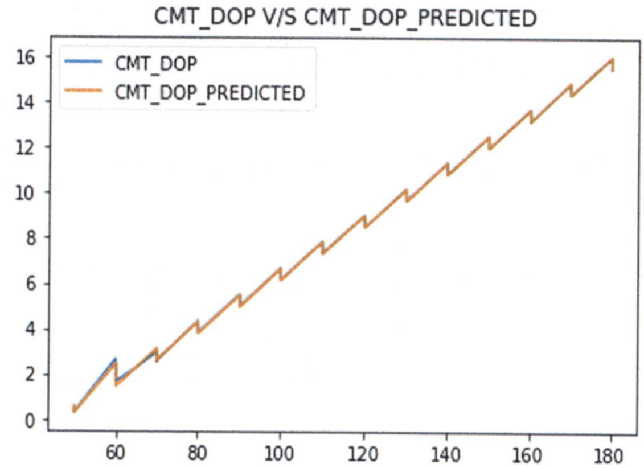

Fig. 2.46 CMT_RH V/S CMT_RH_Predicted

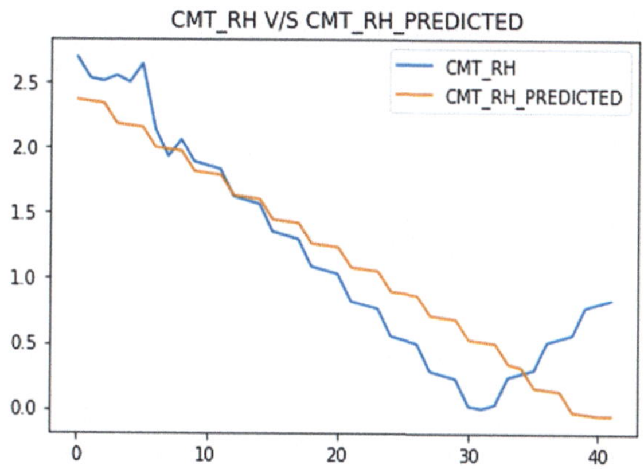

Fig. 2.47 CMT_WPW V/S CMT_WPW_Predicted

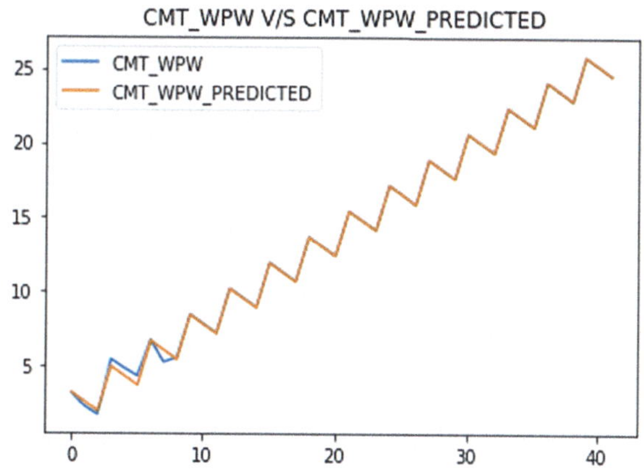

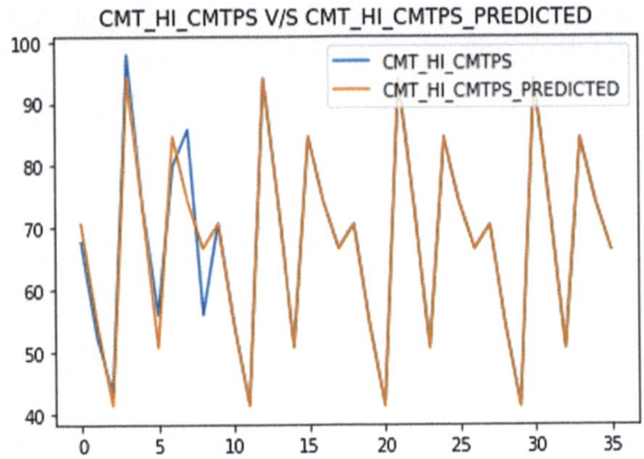

Fig. 2.48 CMT_HI_CMTPS V/S CMT_HI_CMTPS_predicted

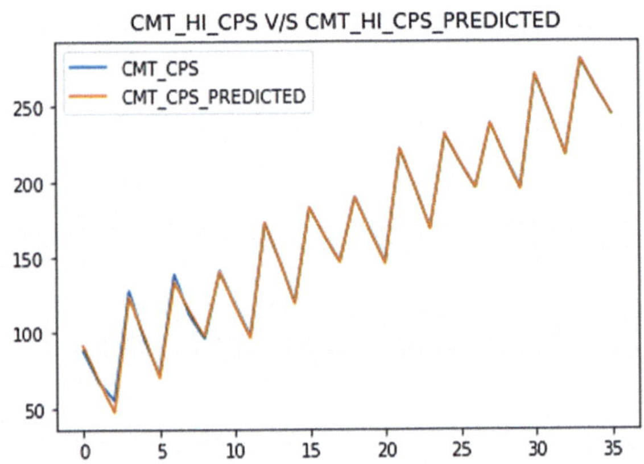

Fig. 2.49 CMT_HI_CPS V/S CMT_HI_CPS_predicted

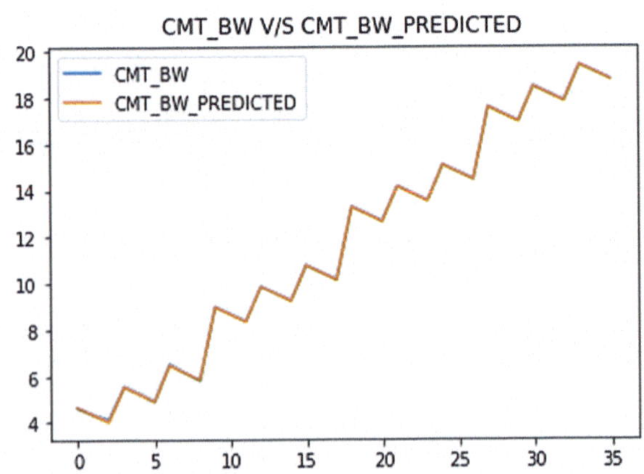

Fig. 2.50 CMT_BW V/S CMT_BW_predicted

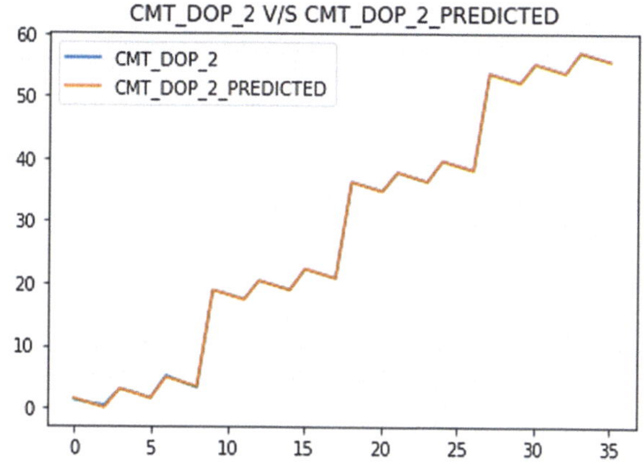

Fig. 2.51 CMT_DOP_2 V/S CMT_DOP_2_predicted

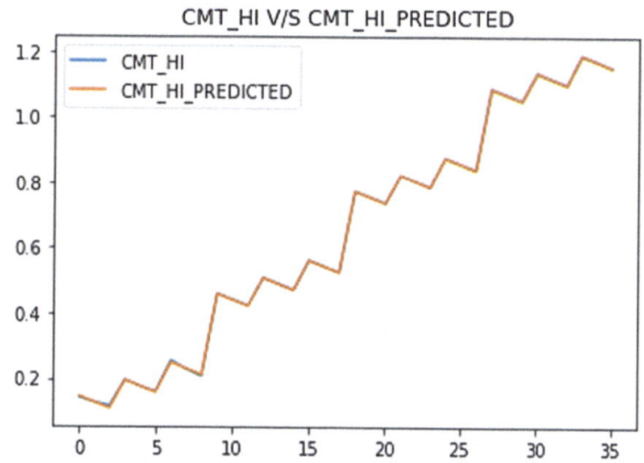

Fig. 2.52 CMT_HI V/S CMT_HI_predicted

From Fig. 2.51, it is observed that the actual value is quite in agreement with the predicted values of CMT_DOP_2.

From Fig. 2.52, it is observed that the actual value is quite in agreement with the predicted values of CMT_HI.

References

1. Kumar NP, Vendan SA, Shanmugam NS (2016) Investigations on the parametric effects of cold metal transfer process on the microstructural aspects in AA6061. J Alloy Comp 658:255–264
2. Pickin CG, Williams SW, Lunt M (2011) Characterisation of the cold metal transfer (CMT) process and its application for low dilution cladding. J Mater Process Technol 211 (3):496–502
3. Pickin CG, Young K (2006) Evaluation of cold metal transfer (CMT) process for welding aluminium alloy. Sci Technol Weld Join 11(5):583–585

4. Zhang HT, Feng JC, He P, Zhang BB, Chen JM, Wang L (2009) The arc characteristics and metal transfer behaviour of cold metal transfer and its use in joining aluminium to zinc-coated steel. J Mat Sci Eng A 499(6):111–113
5. Cao R, Wen BF, Chen JH, Wang PC (2013a) Cold metal transfer joining of magnesium AZ31B-to-aluminum A6061-T6. J Mat Sci Eng A 560:256–266
6. Cao R, Yua G, Chena JH, Wang PC (2013b) Cold metal transfer joining aluminum alloys-to-galvanized mild steel. J Mater Process Technol 213:1753–1763
7. Cao R, Huang Q, Chen JH, Wang PC (2014a) Cold metal transfer spot plug welding of AA6061-T6-to-galvanized steel for automotive applications. J Alloy Comp 585:622–632
8. Cao R, Wang T, Wang C, Feng Z, Lin Q, Chen JH (2014b) Cold metal transfer welding–brazing of pure titanium TA2 to magnesium alloy AZ31B. J Alloy Comp 605:12–20
9. Cao R, Feng Z, Chen JH (2014c) Microstructures and properties of titanium–copper lap welded joints by cold metal transfer technology. Mater Des 53:192–201
10. Cao R, Feng Z, Lin Q, Chen JH (2014d) Study on cold metal transfer welding–brazing of titanium to copper. Mater Des 56:165–173
11. Kang M, Kim C, Kim YM (2016) Joining of AZ31 magnesium alloy and steel sheet under four different coating conditions based on gas metal arc weld-brazing. Mater Trans M2016224
12. Gungor B, Kaluc E, Taban E, Aydin Sik SS (2014) Mechanical and micro-structural properties of robotic cold metal transfer (CMT) welded 5083-H111 and 6082-T651 aluminum alloys. Mater Des 54:207–211
13. Cong B, Ding J, Williams S (2015) Effect of arc mode in cold metal transfer process on porosity of additively manufactured Al-6.3% Cu alloy. Int J Adv Manuf Technol 76(9–12):1593–1606
14. Wu CY, Tung PC, Fuh CC (2010) Development of an automatic arc welding system using an adaptive sliding mode control. J Intell Manuf 21(4):355–362
15. Chen M, Zhang D, Wu C (2017) Current waveform effects on CMT welding of mild steel. J Mater Process Technol 243:395–404
16. Zhang HT, Feng JC, Hu LL (2012) Energy input and metal transfer behavior of CMT welding process [J]. Mater Sci Technol 2
17. Manti R, Dwivedi DK, Agarwal A (2008) Pulse TIG welding of two Al-Mg-Si alloys. J Mater Eng Perform 17:667–673
18. Hu S, Zhang H, Wang Z, Liang Y, Liu Y (2016) The arc characteristics of cold metal transfer welding with AZ31 magnesium alloy wire. J Manuf Process 24:298–306

Chapter 3
Supervised Machine Learning in Friction Stir Welding (FSW)

3.1 Introduction

The friction stir welding process was invented and patented by The Welding Institute (TWI), England in 1991 [1]. FSW process is a solid-state welding technique that is considered as a suitable joining technique for different weld type configurations such as T joints, butt joints, lap joints and butt joints. Currently, FSW process is finding applications in variety of fields such as aerospace, automobile, maritime and railway industries. Frictional stir welding process has several distinct advantages for welding of low melting composites when compared to conventional fusion welding processes. FSW is a clean welding process which does not generate any harmful fumes. By proper selection of machine process parameters, welds free from distortion is achieved and the weld efficiency is also high. Friction stir welding has gained significant attention among the researchers for welding of monolithic materials and metal matrix composites. The application of FSW process has been extensively investigated for flat butt weld joints of composites so far. FSW process can also be successively applied for SiC particle reinforced aluminium metal matrix composites that produce good weldments for reinforcement levels up to 25 weight percentage of SiC particles.

3.2 Process Principle

Friction stir welding appears to address most of the problems in which a customized tool fixed on a probe under rotation travels along the length of the base metal with a face-to-face contact [2, 3]. Interface at the joint between the base metal and tool creates the plastic deformation zone with the stirring action. Subsequently, a thermomechanical plasticized zone is created by the shoulder of the tool and the top surface of the base material and contact among the tool edges and neighbouring

S. A. Vendan et al., *Welding and Cutting Case Studies with Supervised Machine Learning*, Engineering Applications of Computational Methods 1,
https://doi.org/10.1007/978-981-13-9382-2_3

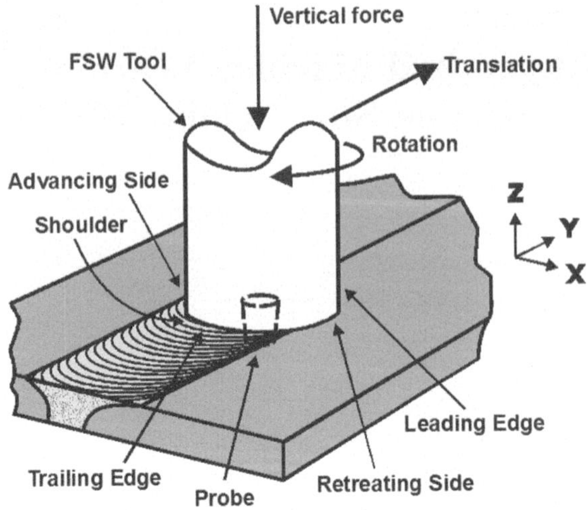

Fig. 3.1 Schematic of FSW process for composite joining [4]

materials instituting plastic deformation. No melting occurs but the material is taken to the state of solidus resulting in a solid-state weld. This has promoted FSW as a prominent player for joining Aluminium Metal Matrix Composites. The FSW process schematic is shown in Fig. 3.1.

3.3 Process-State of Art

This section presents the important methods, analysis techniques, technical problems associated with the welding of composites and significant conclusions drawn by the researchers. The metal matrix composites reveal greater potential to replace conventional materials in case of automobile, marine and other industrial sectors as they demonstrate resistance to fire and radiation, capacity to carry higher heat, greater transverse stiffness and strength, higher electrical and thermal conductivities, no out-gassing and moisture absorption. The other key feature is its higher strength to weight ratio. Important literature reports are discussed under various subsections.

3.3.1 Investigations on FSW Process and the Underlying Physics

This section briefs the experimental investigations reported in literature to understand the process of FSW and the physics governing (Table 3.1).

Table 3.1 Literature reports on FSW process and the underlying physics

Authors and year	Experimentation	Observations	Images and equations
Mishra and Ma [5]	FSW welding of SiC particle reinforced composites	Major process governing parameters in FSW process are tool rotational speed, transverse welding speed and axial force. Other metal parameters include tool material, type of workpiece material, tool geometry	NA
Lee et al. [6].	Investigations of FSW of MMC's	Frequent problem arises with the rate of wear of the tool	NA
Ceschini et al. [7].	Investigations of FSW of MMC's (10 vol.% of Al_2O_3 reinforced AA7005 metal matrix composites)	• There is a substantial grain refinement of the particles in matrix due to dynamic recrystallization caused by frictional heat and intense plastic deformation occurs during welding as observed from microstructural images • Smaller particle fine grains are formed due to stirring and abrasive action made by the tool • Tensile strength results showed a high performance of FSW weld joint, typically about 70–80% of ultimate tensile strength (UTS)	NA
Fernandez and Murr [8]	Investigations on friction stir welding of 20% SiC reinforced AA359 MMC's	• Tool wear profile of the threaded cylindrical steel tools gradually declines with increasing welding speeds and decreasing tool rotational speeds • The threaded tool wears away to a self-optimized shape similar to a pseudo-hour glass at weld traverse distances in surplus of 3 m lower than 10% tool wear • The comparison made among the 25% difference for threaded tools wearing considerably start of the welding and 7% diminution in the SiC mean particle size in the weld region for self-optimized pin tools with no threads	NA

(continued)

Table 3.1 (continued)

Authors and year	Experimentation	Observations	Images and equations
		• The recrystallized grain size diminution and the SiC particles distribution caused the improved hardness of 30% at the weld region compared to parent material	
Xu and Deng [9]	• Investigated the various texture patterns in friction stir welding (FSW) using different cross sections • The text patterns were investigated by using the numerical simulations of the FSW process	• Appearance of bands in the advancing-side material • The onion rings formation was seen in the banded pattern on the traverse cross section. The equal spacing among the bands on the horizontal and longitudinal cross sections with the distance passed through by the welding tool • The texture patterns may be produced due to regular interval spaced material zones occuring at various levels of plastic deformation at the stage of the FSW process	NA
Storjohann et al. [10]	• Investigated the joint behaviours of friction stir welded specimen (A356 alloys) • In particular, about the enhancement of mechanical properties at the weld region was studied for different FSW speeds	• Stir zone (SZ), thermo-mechanically affected zone (TMAZ) and base metal (BM) were observed from the microstructures contained in the weld region • The hypoeutectic Al/Si dendrite structure was displayed by the BM • Even dispersion of the eutectic Si particles was seen in primary Al solid solution • Subsequent to the deformation of the original microstructure, TMAZ is studied in dispersed eutectic Si particles arranged in a line along the rotational direction of the welding tool. These zones are produced at the upper zone and advancing side of the retreating side. • The mechanical properties of the weld region are significantly enhanced in contrast to that of the BM	

(continued)

Table 3.1 (continued)

Authors and year	Experimentation	Observations	Images and equations
		• The degree of formation of Al_4C_3 compounds is directly proportional to the peak weld temperatures	
Rhodes et al. [11], Liu et al. [12]	• Investigated temperature distribution in FSW speciemens	• FSW process tends to plasticize and deform the material around the rotating tool because of frictional heat generated between tooltip and workpieces • Both plasticization and deformation enhance the temperature in and around the stir zone (SZ) • The temperature distribution directly influences the particle grain sizes and distribution, microstructures, coarsening and dissolution of precipitates, and mechanical properties in the stir zone	NA
Mahoney et al. [13]	• Studied various methods for temperature measurements in FSW Process	• Understanding from microstructural behaviour or by employing thermocouple was found to be suitable. • High temperatures and excessive plastic deformation in and around the stir zone results in dynamic recrystallization	NA
Liu [14]	• Investigated FSW joints of fabricated aluminium alloys AA2024 and AA6013	• Tensile results reveal relatively good resistance to the mechanical loadings associated with it • Corrosion resistance testing on AA2024 reveals that the joints are prone to the formation of exfoliation and intergranular corrosion	NA
Nandan et al. [15]	• Investigated the cross sections of FSW of MMC's at the joints	• Observed the fluid flow patterns (Fig. 3.2)	See Fig. 3.2

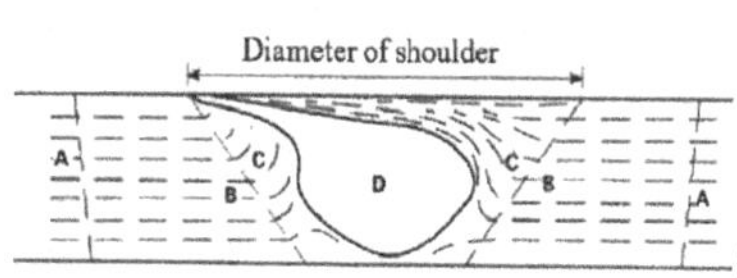

Fig. 3.2 **a** Base Metal **b** Heat Affected Zone **c** Thermo-mechanically affected zone and **d** Stir (nugget) zone

3.3.2 Literature Reports Discussing MATERIAL PARAMETERS Influencing FSW Joints

This section discusses key research reports pertaining to materials composition altering the FSW joint properties (Table 3.2).

Table 3.2 Literature pertaining to the influence of weld materials on weld efficiency and performance

<table>
<tr><th>Authors and year</th><th>Experimentation</th><th>Observations</th><th>Common inference/ images and equations</th></tr>
<tr><td>Mahoney et al. [16]</td><td>• FSW of aluminium metal matrix composites SiC/AA6092</td><td>• FSW was feasible for the composite</td><td rowspan="5">• There is no evidence of occurrence of any chemical reactivity between matrix metal and reinforcements are observed
• Distribution of ceramic reinforcement particles in friction stir welded joints is uniform
• Size and distribution of reinforcement particles in weld nugget zone is similar to that of base composite
• In some samples, it is also observed that the presence of breakdown of reinforcement particles in the weld nugget region and particles are relatively smaller in weld nugget when compared to the base composite</td></tr>
<tr><td>Prado et al. [17], Nakata et al. [18]</td><td>• FSW of aluminium metal matrix composites Al_2O_3/AA6061</td><td>• FSW was feasible for the composite</td></tr>
<tr><td>Nelson et al. [19]</td><td>• FSW of aluminium metal matrix composites B_4C/AA6061</td><td>• FSW was feasible for the composite</td></tr>
<tr><td>Murr et al. [20]</td><td>• FSW of aluminium metal matrix composites SiC/AA339</td><td>• FSW was feasible for the composite</td></tr>
<tr><td>Sharma et al. [21]</td><td>• FSW of aluminium metal matrix composites SiC/AA7093</td><td>• FSW was feasible for the composite</td></tr>
<tr><td>Nandan et al. [15]</td><td>• FSW of aluminium metal matrix composites SiC/AA7093</td><td>• Revealed the presence of more number of SiC particles per given unit area
• However, volume percentage of the SiC particles remains to be same in both the regions (base and nugget)
• This represents the breakdown of larger particles into smaller ones due to FSW tool stirring action</td><td>NA</td></tr>
</table>

(continued)

Table 3.2 (continued)

Authors and year	Experimentation	Observations	Common inference/ images and equations
		• The movement of particles constrained by the grain boundaries allows larger particles knocked down to smaller ones rather than dispersing when subjected under stirring action • The joints exhibit greater mechanical strength and performance over the GMAW welded composite joints. The tensile strength results of FSW composite joints are significantly higher than that of GTA welded composites. Also, yield strength of the FS welded joint is improved	

3.3.3 Literature Reports on FSW Process Parametric Effects on the Weld Joints

This section presents research reports on FSW process parametric effects on welding of aluminium alloys and MMC's (Table 3.3).

Table 3.3 Literature reports pertaining to the process parametric effects on friction stir weldments

Authors and year	Experimentation	Observations	Images and equations
Ceschini et al. [7] and Palanivel et al. [22]	• Effect of tool rotational speed on the tensile strength of FSW joints made of AA6061/20 vol.% Al_2O_3p	• The tensile strength of FS welded AA6351 aluminium alloy joint improves with the increase in tool rotational speed, transverse welding speed and axial force (Figure 3.3) • Ultimate tensile strength (UTM) reaches the maximum value at certain level of process parameters and then gradually decreases	See Fig. 3.3

(continued)

Table 3.3 (continued)

Authors and year	Experimentation	Observations	Images and equations
Kim et al. [23]	• Investigated the sizes of Si reinforcement particles in different weld regions	• The sizes of reinforcement particles remain same for all the tool rotational speeds as shown in Fig. 3.4 • In addition to the welding speed, the tool rotational speed also affects the plasticization and heat energy supplied to the weld region	See Fig. 3.4
Nami et al. [24]	• Investigated friction stir welding of Al/Mg_2Si composites	• Upto a threshold value of rotational speed the tensile strength increaases beyond which it starts declining (Fig. 3.5)	See Fig. 3.5
Storjohann et al. [10]	• Investigated the microstructures of FSW joints of Al-MMC's	• Alignment of SiC whiskers is in the weld line direction	NA
Elangovan et al. [25]	• Investigated the effects of welding speeds on the mechanical strength of FSW of AA60601	• For lower welding speed is inversely proportional to the tensile strength • The tensile strength of AA6061 alloys gradually increases with increasing welding speed until certain level and then joint strength decreases substantially	NA
Cavaliere et al. [26]	• Examined friction stir welded AA6082 joints for their mechanical strength	• Established the relationships between mechanical properties like tensile strength, percentage of elongation, and yield strength and welding travel speed. Figure 3.6 presents the variation of tensile strength with strain%	See Fig. 3.6

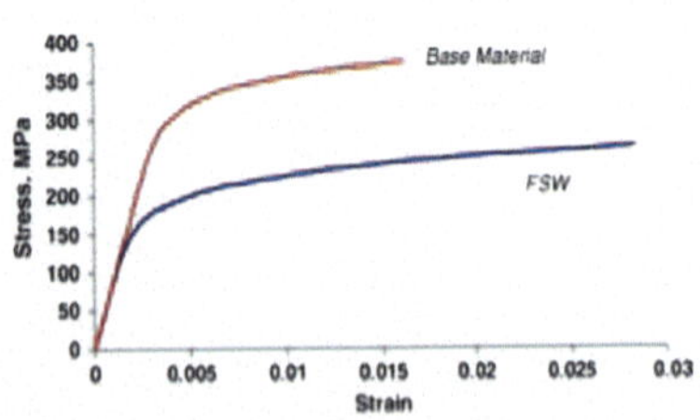

Fig. 3.3 Stress—strain curves for as—received and friction stir welded Al 6061/ Al_2O_3/20p

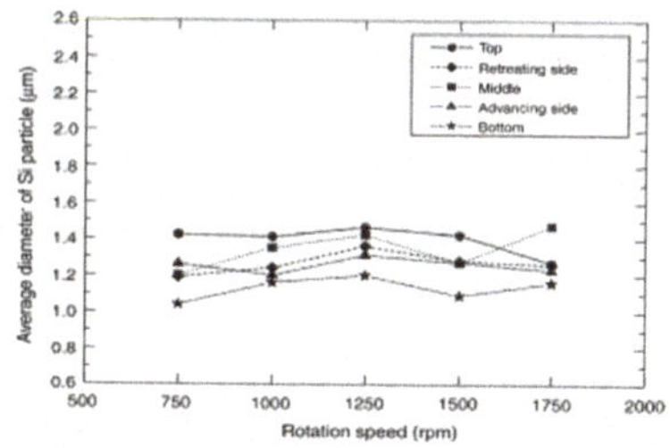

Fig. 3.4 Effect of tool rotational speed on Si particle size

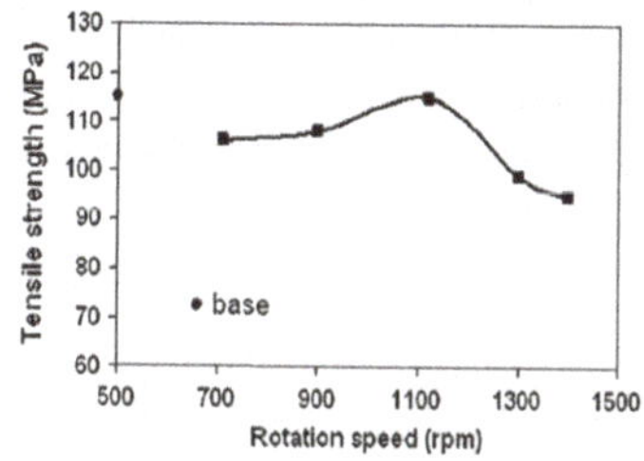

Fig. 3.5 Effects of tool rotational speed on tensile strength of FS welded Al/Mg_2Si joints

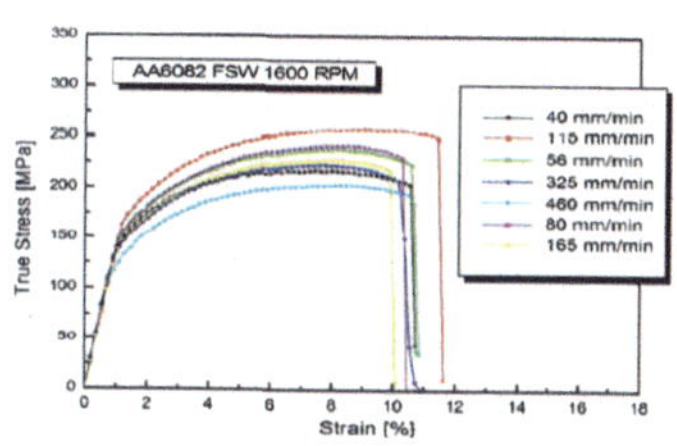

Fig. 3.6 Tensile properties of the studied joints revealing the variation of the weld strength by varying the welding speed

3.4 Experimentation

3.4.1 Material

Aluminium alloy, AA6061 is used as the matrix metal. Aluminium is widely researched material because of its inherent advantages [27]. Table 3.4 shows the chemical composition of the AA6061.

Table 3.4 Chemical composition of Aluminium Alloy (AA6061) [28]

Component	Al	Mg	Si	Fe	Cu	Zn	Ti	Mn	Cr	Others
Amount (wt%)	Bal.	0.8–1.2	0.4–0.8	0.7	0.15–0.40	0.25	0.15	0.15	0.04–0.35	0.05

3.4.2 Reinforcement Particles

The present research study adopted TiB_2, SiC and B_4C reinforcement particulates in various compositions and aluminium alloy, AA6061 as matrix metal.

3.4.3 Preparation, Melting and Casting

The metal matrix composites of AA6061 reinforced with different weight percentages of SiC, B_4C and TiB_2 particulates are shown in Table 3.5 and they were prepared in a crucible furnace by employing stir casting. Figure 3.7 presents the FSW welded samples with few portions sectioned to carry out mechanical testing and metallurgical characterization.

Table 3.5 Reinforcement particles weight percentage for fabricated composites [28]

Samples	Phase I				Phase II			
	Al %	SiC %	B_4C %	Mg %	Al %	SiC %	TiB_2 %	Mg %
1	75	12.5	7.5	2	80	9	6	5
2	75	10.5	12.5	2	80	7	8	5
3	75	13	10	2	80	5	10	5
4	75	5.5	17.5	2	80	3	12	5

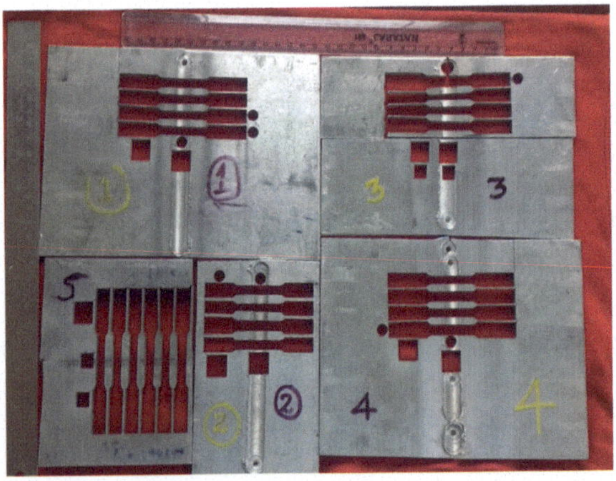

Fig. 3.7 FSW welded composites plates [28]

Table 3.6 Process parameters for machine [28]

Parameter	Value
Tool rotational speed (rpm)	600
	800
	1000
Transverse speed (mm/min)	50
	60
	70

3.4.4 FSW Machine Setup

AA6061/TiB_2/SiC metal matrix composite with different weight percentages of SiC, B_4C and TiB_2 plates with 140 × 120 × 8 mm dimensions were prepared from casting. The machine process parameters such as tool rotating speed, travelling speed and plunge force were shown in Table 3.6. After series of trial and error as well as design of experiments based runs, the optimal welding parameters are carefully chosen to achieve a defect-free weld.

3.4.5 Process Parameters

FSW process has two significant process parameters: 1. Tool rotation speed and 2. Transverse speed. The range of these two parameters are varied during experimentation is shown in Table 3.4. The process parameters mainly classified into dependent and independent properties. The dependent parameters are the outcomes of independent parameters, includes temperature gradient, residual stresses and heat generation. The independent parameters include tool rotation rate, transverse speed, pressure, tool dimensions and tilt angle. High value of tool rotation speeds, i.e. increased friction action results in severe plastic deformation and excessive heat generation. Lower tool rotating speeds tend to produce small and homogeneous grains within the weld region. Higher transverse speed results in less heat applied per unit area, whereas in case of lower speeds, more stirring action accounts at the weld region. High tool rotation speed, low transverse speeds and high axial pressure tend to increase the temperature distribution profile.

3.4.6 FSW Tool Geometry

In the present study, cylindrical tool material used for the FSW process is H-13 tool steel oil hardened to 60 HRC. The FSW tool was manufactured with shoulder 24 mm in diameter, the probe tip diameter is of 8 mm as shown in Fig. 3.8.

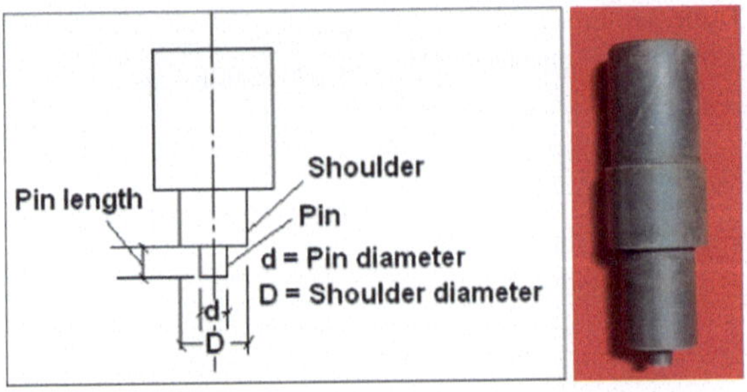

Fig. 3.8 Friction stir welding tool [28]

3.4.7 *Estimation of Heat Generated During Friction Stir*

Power consumption of the machine was monitored while welding of the different metals. The power–time characteristics were employed frequently as the indication of heat flux for analysis and modelling of friction stir welding process. Therefore, the experimentally measured power data is used to define the heat flux boundary condition at the weld interface, as the third variant of the heat-generation estimate. The plot of recorded/experimental power consumed during friction stir welding was shown in Fig. 3.9. The curve shows that the power dissipation raises at the initial peak value. This is the required power in order to reach a predetermined rotational speed. At this stage, the tool and the workpiece are not in contact, and therefore it is included in the calculation of the heat-generation rate. Then, the power increases to its peak value as the frictional force is exerted. An efficiency factor of 7–10% is defined for the machine because the recorded power is not entirely converted to heat at the friction interface. Table 3.7 presents the heat input values for FSW metal matrix composites [28].

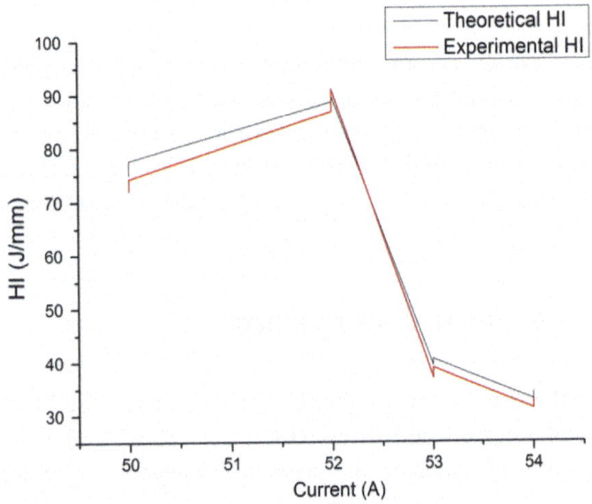

Fig. 3.9 Effect of welding current on theoretical and experimental heat input values [28]

Table 3.7 Heat input values for friction stir welded metal matrix composites [28]

S. no	Welding current (A)	Welding voltage (V)	Theoretical heat input (Q_t) (J/mm)	Torque developed (Nm/mm)	Experimental heat generated (Q_e) (J/mm)
1	50	150	75	1.12	72.12
2	50	155	77.64	1.24	74.28
3	52	160	88.6	1.5	86.8
4	52	165	89.65	1.56	90.96
5	53	170	39.33	1.66	37.01
6	53	175	40.56	1.72	38.96
7	54	180	33	1.98	31.31
8	54	185	34.35	2.01	32.85

Based on power calculations, the input energy supplied to the friction welding is

$$E = VIt\,\text{Joules} \tag{3.1}$$

V, *I* are the welding voltage and welding current, respectively, and time t is the rotation time of the tool.

According to ASME Section IXQW-409.1, theoretical heat input equation for friction stir welding process is as follows:

$$Q_t = \frac{Voltage \times Amperage \times 60}{Travel\,speed(\text{mm/min})}\text{J/mm} \tag{3.2}$$

The correlation between the heat generated and torque is expressed as

$$dQ = \omega.dM \tag{3.3}$$

Torque generated during the friction stir welding is

$$dM = r.dF_{\text{friction}} \tag{3.4}$$

$$dF_{\text{friction}} = r(\mu.P.2\pi r.\text{d}r) \tag{3.5}$$

We obtain heat occurred at welding surface by integrating with R of Eq. (3.5) as follows:

$$Q = \frac{2}{3}\pi P\omega R^3 \tag{3.6}$$

Q—Heat generated, M—Torque developed, ω—Rotational speed, F_{friction}—Frictional force, R—Radius of tool.

With the increase in welding current, there is an increased molten metal drop rate results in more amount of generation of heat, which decreases with increase in welding speed while maintaining constant welding current. This results in the less deposition of material per unit length and less driving force for molten weld pool to spread on the surface of the substrate. However, from Fig. 3.9 it is clear that the heat input values were in non-linear behaviour with the welding current.

The heat input values are observed to decrease with increasing speed.

3.5 Results and Discussion

3.5.1 Tensile Strength

The UTS of the fabricated AA6061/SiC/B_4C composite was estimated from the test specimens prepared as per ASTM E8-04 as shown in Fig. 3.10 obtained from each composite. In Table 3.8, UTS is the ultimate tensile strength, YS is the yield strength (determined at 0.2% of plastic deformation) and *E* is the modulus of elasticity. Selection of process parameters is very important in FSW process, improper selection often leads to the discontinuities in the weld zone and results in initiation of cracks. The weld efficiency of 92% has been achieved for the process parameter setting of tool rotational speed—1000 rpm and transverse welding speed —70 mm/min. The welded joint strength increases with increasing tool rotational speed from 600 rpm to 1000 rpm and decreases with further increase in tool rotational speeds. Excessive heat generation at the weld joint was not preferable; excessive stirring action results in deprived mechanical properties. It was found that the tool rotational speed is major governing factor when compared to other machine and metal parameters. The less heat generation takes place for lower tool rotational speeds, results in reduction of heat transfer from FSW tool to the weld region. Low heats at the weld region lead to insufficient material flow and less plasticization of the material in nugget zone. Therefore, low magnitudes of tensile strength observed for lower tool rotational speeds [29]. When tool rotational speed is very high

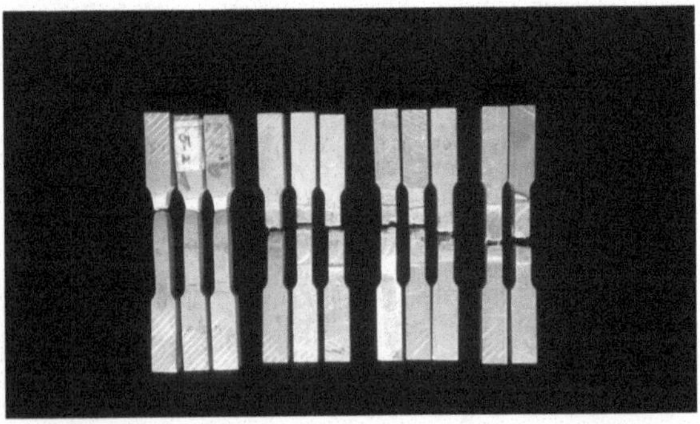

Fig. 3.10 Tensile test specimen prepared from friction stir welded AA6061/SiC/B_4C composite [28]

Table 3.8 Tensile strength of AA6061/SiC/B_4C composite [28]

S. no.	Transverse speed (mm/min)	Tool rotational speed (rpm)	UTS (MPa)	YS (MPa)	ε%	Failure location
1	50	600	116	95	22.267	SZ
2	60	600	98	89	15.334	TMAZ
3	70	600	104	91	12.645	SZ
4	50	800	95	82	19.569	TMAZ
5	60	800	106	92	17.258	BM
6	70	800	112	94	16.235	BM
7	50	1000	85	82	21.232	SZ
8	60	1000	97	96	18.245	TMAZ
9	70	1000	124	95	16.558	SZ

(seldom not required), for higher tool speeds, heat generated is also higher, which results in more plasticization of the material and the stirring action causes turbulence effects in the material flow. The combination of these two effects, namely, plasticization and stirring effects led to the grain growth. Thus, lower tensile strengths are obtained for higher tool rotational speeds [29]. The fractures occurred during tensile testing are confirmed as ductile fractures, surfaces of fracture specimens have tear ridges and dimples to support the observations. At the bottom end of dimples, SiC and B_4C particles were observed. The particles inside the fracture voids suggest the origin of crack initiation in parent matrix material, i.e. AA6061. In addition to this, the dendritic whiskers at the fracture surface show that propagation of fracture happens in inter-dendritic separation. In Fig. 3.10, for particular set of specimens, the failure location observed at weld nugget zones and crack growth phenomena observed in the weld direction. The failed specimens at the weld region are for higher welding speeds and tool rotational speeds.

Three distinct failure zones are observed during the tensile testing, viz., base metal (BM), weld zone (WZ), heat-affected zone and in the thermo-mechanically affected zone (TMAZ). Failure at TMAZ occurs on the advancing side (AS) of the weld. The fractures observed in the stirred zone are mainly due to localized defects such as root flaws and lack of penetration. Joint efficiency represents the ultimate tensile strength of the welded joint ratio while one of the base materials represents an indicator of the weld strength. The reinforcement particles have a significant influence on the strength of the MMCs. For the base material (BM), an increased concentration of B_4C particles increases the strength and reduces the ductility of the composite. The welded joint efficiency was limited to about 60%.

3.5.2 *Metallurgical Properties of Friction Stir Welded AA6061/Sic/B_4C Composites*

During the friction stir welding process of the metal matrix composites, due to variation in thermal cycles, there is significant change in the weld region microstructures. Such variation in microstructures sometimes might be dependent on the process parameters to a greater extent, which in turn alter the joint properties. Studies on these aspects require detailed understanding of the metallurgical behaviour with respect to process and metal parameters.

3.5.2.1 Microstructures

The microstructure of the AA6061/SiC/B_4C composite is thoroughly examined and discussed in this section. Typically, the microstructure reveals a uniform distribution of the B_4C/SiC reinforcement in the AA6061 alloy matrix. The faying surfaces of the two AA6061/SiC/B_4C composite plates were successfully welded by friction stir welding.

The microstructures of welded composites are shown in Fig. 3.11. The microstructure of FSW joints can be separated into four zones: (1) parent material; (2) heat-affected zone (HAZ); (3) thermo-mechanical affected zone (TMAZ) and (4) weld nugget. The thermo-mechanically affected zone (TMAZ), which is adjacent to the weld nugget, i.e. Retreating side or the advancing side, has undergone plastic deformation and are thermally affected. TMAZ is illustrated by a rotation of up to 90° of both the elongated grains of the Al alloy matrix and the reinforced particle-free regions of the composite. The reinforced particle alignment was also observed in the TMAZ region. The heat-affected zone (HAZ) between TMAZ and unaffected base composite regions both at the retreating and advancing sides exhibit a microstructure similar to the base metal.

The microstructure at the joint exhibits a continuous flow of plasticized material within the advancing and retreating side as shown in Fig. 3.11. Onion rings are formed owing to the effect of geometry on the cylindrical sheets of material extruded in the process of tool rotation that cuts the material portion creating a well-defined 'Onion Rings'. Typical, FSW defects such as piping, tunnels and wormholes are avoided owing to the presence of B_4C which resists the free flow of matrix alloy during FSW. The weld nugget structure typically develops onion rings in the AMCs but similar observation is eluded in the weld nugget of friction stir welded alloys. During the joining process, re-arrangement of the particles occurs due to high deformation and stirring. The stirring action made by the FSW tool refines the grains at the weld region centre. Due to plastic deformation at the weld region the grains becomes finer. The microstructures at the nugget zone significantly vary from the other regions such as base metal and heat-affected zones. Macroscopic image is shown in Fig. 3.12.

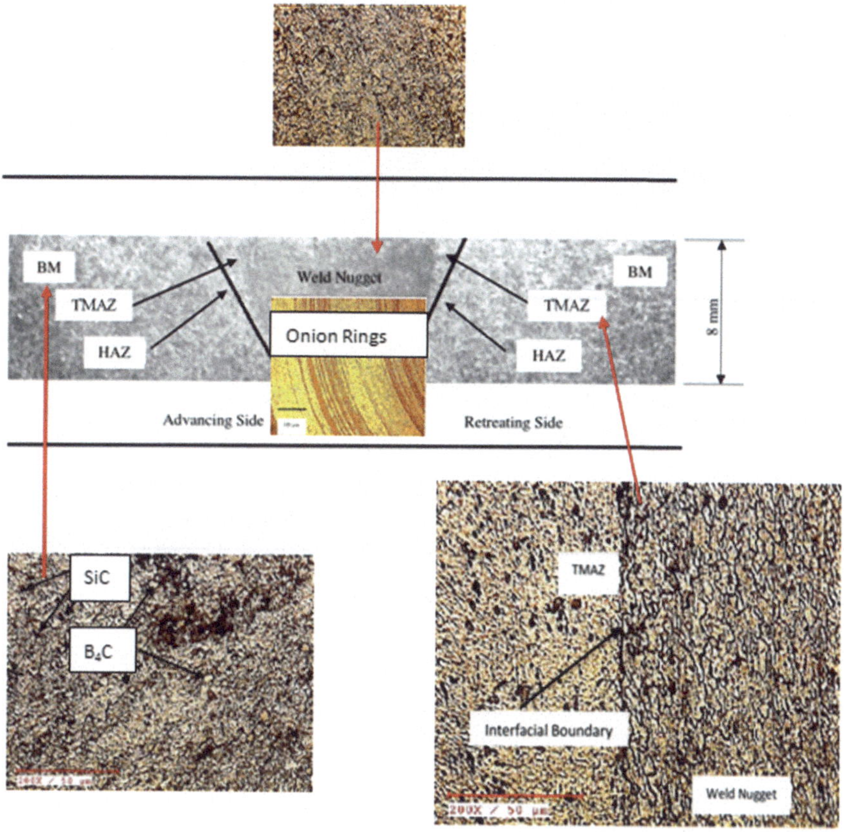

Fig. 3.11 Microstructure zones of the FSW welded composite samples. **a** Base material **b** HAZ **c** TMAZ and **d** weld region [28]

The changes in weld region macrostructures and microstructures mainly are attributed to the plastic deformation and frictional heat generated due to FSW stirring tool [30]. High strain created by the stirring action in the weld zone breaks and rearranges the SiC and B_4C reinforcement particulates and precipitates. Agglomeration of the particles at the grain boundary observed in the base material and homogeneous distribution of particles were observed in the nugget zone of welded composite. The smaller and finer equiaxed grains formed in the nugget zone is the outcome of consistent dynamic recrystallization [30, 31]. In thermo-mechanically affected zone (TMAZ), distributions of highly elongated

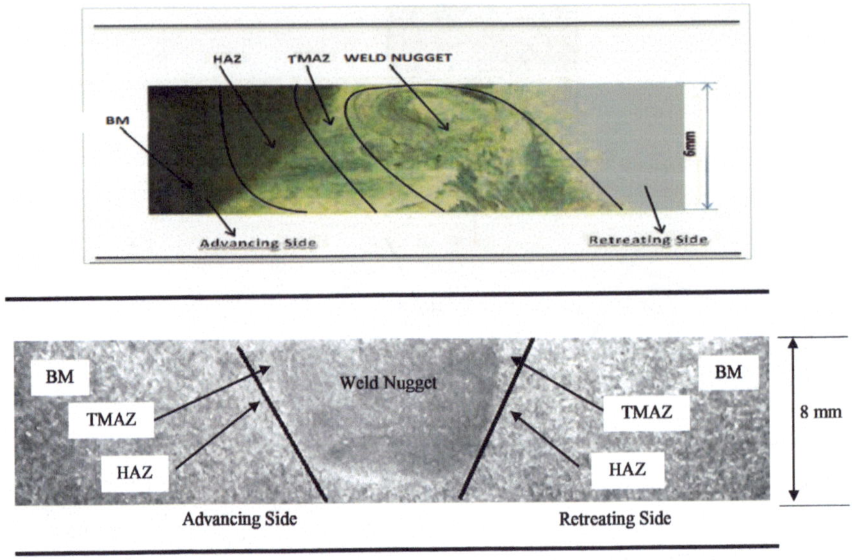

Fig. 3.12 Macroscopic view at the cross section of the friction stir welded AA6061/SiC/B_4C composite illustrating typical weld zones (Weld nugget, TMAZ, HAZ) [28]

grains along the grain boundary have been observed [32, 33]. This elongation of grains due to insufficient recrystallization is the result in combination of heat and plastic deformation [29, 34, 35]. In heat-affected zone the grains experience only heat and not the plastic deformation.

3.5.2.2 Scanning Electron Microscopy for Analysis

SEM images of weld specimens reveal that B_4C and SiC particles are uniformly and homogenously distributed in the metal matrix. No segregation is observed at any particular region as shown in Fig. 3.13. There are traces of porosity at tips of B_4C as observed from the SEM images. There is no fragmentation or rounding of individual B_4C particles during stirring for welded specimens with B_4C (Fig. 3.14). Two influential parameters that affect the strength of the joint are the strength of the reinforcement particles and plastic deformation due to stirring.

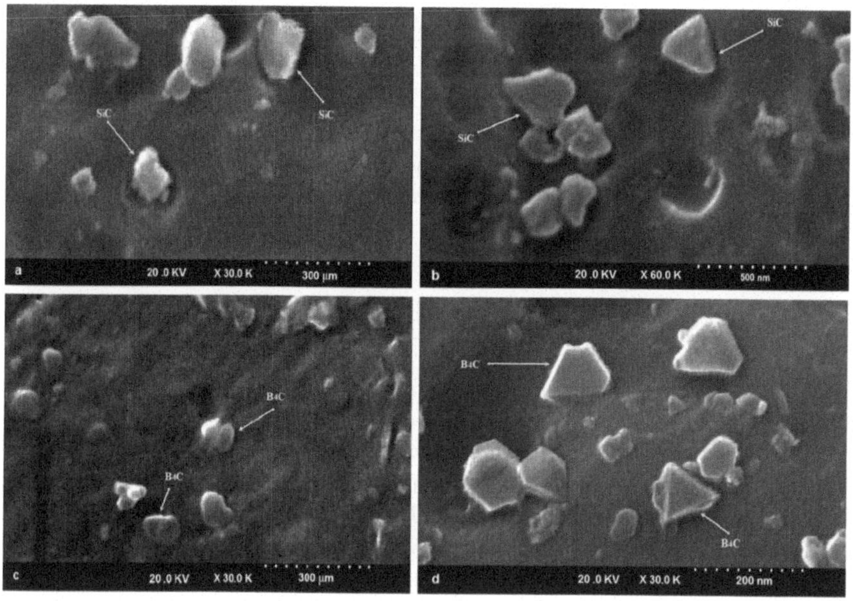

Fig. 3.13 SiC and B_4C particles in aluminium matrix composites [28]

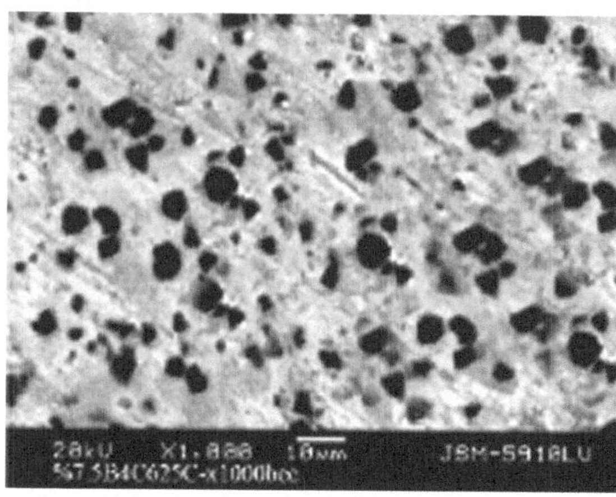

Fig. 3.14 SEM images of B_4C particles in aluminium metal matrix for weight percentages of 7.5 wt% B_4C for Al matrix [28]

3.6 Parametric Analysis Using Machine Learning Terminologies

This section presents details on FSW data analysis and predictions using machine learning techniques. The fundamental concepts based on which this work is performed has been discussed in Chap. 1 (Sect. 1.5).

3.6.1 Data Analysis Using NumPy *and* Pandas

Since the data structures are going to be frequently used, importing them into the local namespace is equally important and can be achieved in the following way.

Understanding pandas requires fundamental prerequisites about its widely used data structures such as Series and DataFrame. Now, let us individually discuss each of these data structures (Figs. 3.15 and 3.16).

The first step is to import the dataset. Figures 3.17, 3.18, 3.19, 3.20, 3.21, 3.22, 3.23 and 3.24 give an idea on how to import the dataset using pandas (Fig. 3.25).

3.6.2 Data Visualization Using Seaborn *and* Matplotlib

The analysis and visualization done for the datasets are to find out the co-relation between the independent variables, i.e. inputs and dependent variable, i.e. output by the help of seaborn library. Figures 3.26, 3.27, 3.28, 3.29, 3.30, 3.31, 3.32 and 3.33 show the heat map and co-relation matrix for each dataset.

Fig. 3.15 Instructions to import Pandas

```
import pandas as pd
```

Fig. 3.16 Instructions to import series and DataFrame from Pandas

```
from pandas import Series, DataFrame
```

Fig. 3.17 Instructions to import FSW_TS Dataset

```
# Load the data and replace the '..' with nan
ts_df = pd.read_excel('FSW_TS.xls')

ts_df = ts_df.astype(float)

display(ts_df)
```

	Rotating Speed(RPM)	Welding Speed(mm/min)	axial force(Kn)	tensile strength(mpa)
0	900.0	40.0	3.0	323.090000
1	900.0	40.0	5.0	311.020000
2	900.0	40.0	7.0	308.160000
3	900.0	50.0	3.0	299.010000
4	900.0	50.0	5.0	306.160000
5	900.0	50.0	7.0	289.620000
6	900.0	60.0	3.0	306.990000
7	900.0	60.0	5.0	307.370000
8	900.0	60.0	7.0	352.350000
9	1100.0	40.0	3.0	319.980000
10	1100.0	40.0	5.0	274.560000
11	1100.0	40.0	7.0	336.160000
12	1100.0	50.0	3.0	300.610000
13	1100.0	50.0	5.0	310.920000
14	1100.0	50.0	7.0	345.980000
15	1100.0	60.0	3.0	270.490000
16	1100.0	60.0	5.0	358.430000
17	1100.0	60.0	7.0	326.790000
18	1300.0	40.0	3.0	328.880000
19	1300.0	40.0	5.0	277.410000
20	1300.0	40.0	7.0	288.480000
21	1300.0	50.0	3.0	262.020000
22	1300.0	50.0	5.0	285.520000
23	1300.0	50.0	7.0	298.680000
24	1300.0	60.0	3.0	217.480000
25	1300.0	60.0	5.0	275.420000
26	1300.0	60.0	7.0	299.390000
27	1500.0	40.0	3.0	263.831667
28	1500.0	40.0	5.0	275.890556
29	1500.0	40.0	7.0	287.949444
30	1500.0	50.0	3.0	260.885556
31	1500.0	50.0	5.0	272.944444
32	1500.0	50.0	7.0	285.003333
33	1500.0	60.0	3.0	257.939444
34	1500.0	60.0	5.0	269.998333
35	1500.0	60.0	7.0	282.057222
36	1700.0	40.0	3.0	248.804444
37	1700.0	40.0	5.0	260.863333
38	1700.0	40.0	7.0	272.922222
39	1700.0	50.0	3.0	245.858333
40	1700.0	50.0	5.0	257.917222
41	1700.0	50.0	7.0	269.976111
42	1700.0	60.0	3.0	242.912222
43	1700.0	60.0	5.0	254.971111
44	1700.0	60.0	7.0	267.030000

```
# Load the data and replace the '..' with nan
ts_df = pd.read_excel('FSW_EHG.xlsx')

ts_df = ts_df.astype(float)

display(ts_df)
```

	Welding Current (A)	Welding Voltage (V)	Experimental Heat generated (Qe) (J/mm)
0	50.0	150.0	72.12
1	50.0	155.0	74.28
2	52.0	160.0	86.80
3	52.0	165.0	90.96
4	53.0	170.0	37.01
5	53.0	175.0	38.96
6	54.0	180.0	31.31
7	54.0	185.0	32.85

Fig. 3.18 Instructions to import FSW_EHG

From the co-relation matrix of FSW_TS mentioned above, the following input parameters such as WELDING SPEED and ROTATING SPEED are NEGETAVILY co-related to TS, i.e. Tensile Strength.

From the co-relation matrix of FSW_EHG mentioned above, the following input parameters such as WELDING VOLTAGE and WELDING CURRENT are negatively co-related to EHG, i.e. EXPERIMENTAL HEAT GENERATED, respectively.

From the co-relation matrix of FSW_TD mentioned above, the following input parameters such as WELDING VOLTAGE and WELDING CURRENT are positively co-related to TD, i.e. TORQUE DEVELOPED.

From the co-relation matrix of FSW_THI mentioned above, the following input parameters such as WELDING VOLTAGE and WELDING CURRENT are positively co-related to THI, i.e. THEORETICAL HEAT INPUT.

From the co-relation matrix of FSW_EL mentioned above, the following input parameter such as TOOL ROTATION SPEED is positively co-related to Ɛ%.

From the co-relation matrix of FSW_EL mentioned above, the following input parameter such as TOOL ROTATION SPEED is negatively co-related to UTS.

From the co-relation matrix of FSW_YL mentioned above, the following input parameter such as TRANSVERSE SPEED is positively co-related to UTS.

Fig. 3.19 Instructions to import FSW_TD

```
# Load the data and replace the '..' with nan
ts_df = pd.read_excel('FSW_TD.xlsx')

ts_df = ts_df.astype(float)

display(ts_df)
```

	Welding Current (A)	Welding Voltage (V)	Torque Developed (Nm/mm)
0	50.0	150.0	1.120000
1	50.0	155.0	1.240000
2	52.0	160.0	1.500000
3	52.0	165.0	1.560000
4	53.0	170.0	1.660000
5	53.0	175.0	1.720000
6	54.0	180.0	1.980000
7	54.0	185.0	2.010000
8	55.0	190.0	2.168030
9	55.0	195.0	2.236377
10	56.0	200.0	2.399894
11	56.0	205.0	2.468242
12	57.0	210.0	2.631758
13	57.0	215.0	2.700106
14	58.0	220.0	2.863623
15	58.0	225.0	2.931970
16	59.0	230.0	3.095487
17	59.0	235.0	3.163835
18	60.0	240.0	3.327352
19	60.0	245.0	3.395699
20	61.0	250.0	3.559216
21	61.0	255.0	3.627564
22	62.0	260.0	3.791081
23	62.0	265.0	3.859428
24	63.0	270.0	4.022945
25	63.0	275.0	4.091292
26	64.0	280.0	4.254809
27	64.0	285.0	4.323157
28	65.0	290.0	4.486674
29	65.0	295.0	4.555021
30	66.0	300.0	4.718538
31	66.0	305.0	4.786886

Fig. 3.20 Instructions to import FSW_THI

```
# Load the data and replace the '..' with nan
ts_df = pd.read_excel('FSW_THI.xlsx')

ts_df = ts_df.astype(float)

display(ts_df)
```

	Welding Current (A)	Welding Voltage (V)	Theoretical Heat Input (Qt) (J/mm)
0	50.0	150.0	75.00
1	50.0	155.0	77.64
2	52.0	160.0	88.60
3	52.0	165.0	89.65
4	53.0	170.0	39.33
5	53.0	175.0	40.56
6	54.0	180.0	33.00
7	54.0	185.0	34.35

From the co-relation matrix of FSW_W mentioned above, the following input parameters such as TIME and FRICTIONAL FORCE are positively co-related to W.

3.6.3 Data Preprocessing Using Scikit-Learn

Data preprocessing is one of the crucial step in machine learning. This is an unavoidable step incase of data which are in unstructured form. Further to this, we will be discussing data preprocessing by applying machine learning libraries such as Scikit-Learn. This is a preprocessing technique used to derive non-linear features. It is mostly applied along with linear regression algorithm to train the model of higher degree. It is implemented in the following manner.

Fig. 3.21 Instructions to import FSW_EL

```
# Load the data and replace the '..' with nan
ts_df = pd.read_excel('FSW_EL.xlsx')

ts_df = ts_df.astype(float)

display(ts_df)
```

	Transverse speed (mm/min)	Tool rotational speed (rpm)	ε%
0	50.0	600.0	22.267000
1	60.0	600.0	15.334000
2	70.0	600.0	12.645000
3	50.0	800.0	19.569000
4	60.0	800.0	17.258000
5	70.0	800.0	16.235000
6	50.0	1000.0	21.232000
7	60.0	1000.0	18.245000
8	70.0	1000.0	16.558000
9	50.0	1200.0	22.572778
10	60.0	1200.0	19.634444
11	70.0	1200.0	16.696111
12	50.0	1400.0	23.537611
13	60.0	1400.0	20.599278
14	70.0	1400.0	17.660944
15	50.0	1600.0	24.502444
16	60.0	1600.0	21.564111
17	70.0	1600.0	18.625778
18	50.0	1800.0	25.467278
19	60.0	1800.0	22.528944
20	70.0	1800.0	19.590611
21	50.0	2000.0	26.432111
22	60.0	2000.0	23.493778
23	70.0	2000.0	20.555444
24	50.0	2200.0	27.396944
25	60.0	2200.0	24.458611
26	70.0	2200.0	21.520278
27	50.0	2400.0	28.361778
28	60.0	2400.0	25.423444
29	70.0	2400.0	22.485111
30	50.0	2600.0	29.326611
31	60.0	2600.0	26.388278
32	70.0	2600.0	23.449944
33	50.0	2800.0	30.291444
34	60.0	2800.0	27.353111
35	70.0	2800.0	24.414778
36	50.0	3000.0	31.256278
37	60.0	3000.0	28.317944
38	70.0	3000.0	25.379611
39	50.0	3200.0	32.221111
40	60.0	3200.0	29.282778
41	70.0	3200.0	26.344444
42	50.0	3400.0	33.185944
43	60.0	3400.0	30.247611
44	70.0	3400.0	27.309278

Fig. 3.22 Instructions to import FSW_UTS

```
# Load the data and replace the '..' with nan
ts_df = pd.read_excel('FSW_UTS.xlsx')

ts_df = ts_df.astype(float)

display(ts_df)
```

	Transverse speed (mm/min)	Tool rotational speed (rpm)	UTS(MPa)
0	50.0	600.0	116.000000
1	60.0	600.0	98.000000
2	70.0	600.0	104.000000
3	50.0	800.0	95.000000
4	60.0	800.0	106.000000
5	70.0	800.0	112.000000
6	50.0	1000.0	85.000000
7	60.0	1000.0	97.000000
8	70.0	1000.0	124.000000
9	50.0	1200.0	92.777778
10	60.0	1200.0	100.111111
11	70.0	1200.0	107.444444
12	50.0	1400.0	90.777778
13	60.0	1400.0	98.111111
14	70.0	1400.0	105.444444
15	50.0	1600.0	88.777778
16	60.0	1600.0	96.111111
17	70.0	1600.0	103.444444
18	50.0	1800.0	86.777778
19	60.0	1800.0	94.111111
20	70.0	1800.0	101.444444
21	50.0	2000.0	84.777778
22	60.0	2000.0	92.111111
23	70.0	2000.0	99.444444
24	50.0	2200.0	82.777778
25	60.0	2200.0	90.111111
26	70.0	2200.0	97.444444
27	50.0	2400.0	80.777778
28	60.0	2400.0	88.111111
29	70.0	2400.0	95.444444
30	50.0	2600.0	78.777778
31	60.0	2600.0	86.111111
32	70.0	2600.0	93.444444
33	50.0	2800.0	76.777778
34	60.0	2800.0	84.111111
35	70.0	2800.0	91.444444

Fig. 3.23 Instructions to importing FSW_YS

```
# Load the data and replace the '..' with nan
ts_df = pd.read_excel('FSW_YS.xlsx')

ts_df = ts_df.astype(float)

display(ts_df)
```

	Transverse speed (mm/min)	Tool rotational speed (rpm)	YS(MPa)
0	50.0	600.0	95.000000
1	60.0	600.0	89.000000
2	70.0	600.0	91.000000
3	50.0	800.0	82.000000
4	60.0	800.0	92.000000
5	70.0	800.0	94.000000
6	50.0	1000.0	82.000000
7	60.0	1000.0	96.000000
8	70.0	1000.0	95.000000
9	50.0	1200.0	86.500000
10	60.0	1200.0	90.000000
11	70.0	1200.0	93.500000
12	50.0	1400.0	86.166667
13	60.0	1400.0	89.666667
14	70.0	1400.0	93.166667
15	50.0	1600.0	85.833333
16	60.0	1600.0	89.333333
17	70.0	1600.0	92.833333
18	50.0	1800.0	85.500000
19	60.0	1800.0	89.000000
20	70.0	1800.0	92.500000
21	50.0	2000.0	85.166667
22	60.0	2000.0	88.666667
23	70.0	2000.0	92.166667
24	50.0	2200.0	84.833333
25	60.0	2200.0	88.333333
26	70.0	2200.0	91.833333
27	50.0	2400.0	84.500000
28	60.0	2400.0	88.000000
29	70.0	2400.0	91.500000
30	50.0	2600.0	84.166667
31	60.0	2600.0	87.666667
32	70.0	2600.0	91.166667
33	50.0	2800.0	83.833333
34	60.0	2800.0	87.333333
35	70.0	2800.0	90.833333

Fig. 3.24 Instructions to import FSW_W

```
# Load the data and replace the '..' with nan
ts_df = pd.read_excel('FSW_W.xlsX')

ts_df = ts_df.astype(float)

display(ts_df)
```

	Time (s)	Frictional Force (N)	Wear (µm)
0	30.386	3.02	4.13
1	31.366	3.06	4.15
2	32.348	3.07	4.17
3	33.328	3.08	4.11
4	34.307	3.07	4.04
5	35.288	3.07	3.97
6	36.269	3.07	3.91
7	37.248	3.07	3.92
8	38.229	3.07	3.97
9	39.208	3.07	4.05
10	40.189	3.09	4.18
11	41.169	3.11	4.36
12	42.150	3.12	4.62
13	43.130	3.13	4.92
14	44.109	3.14	5.23
15	45.089	3.15	5.54
16	46.070	3.17	5.81
17	47.050	3.19	6.08
18	48.030	3.21	6.29
19	49.010	3.22	6.40
20	49.992	3.24	6.42

```
import seaborn as sns
import pandas as pd
import matplotlib.pyplot as plt
```

Fig. 3.25 Instructions to import libraries for data visualization and analysis

```
# calculate the correlation matrix
corr = ts_df.corr()
# display the correlation matrix
display(corr)
# plot the correlation heatmap
sns.heatmap(corr, xticklabels=corr.columns, yticklabels=corr.columns, cmap='RdBu')
```

	Rotating Speed(RPM)	Welding Speed(mm/min)	axial force(Kn)	tensile strength(mpa)
Rotating Speed(RPM)	1.000000	0.000000	0.000000	-0.698022
Welding Speed(mm/min)	0.000000	1.000000	0.000000	-0.079009
axial force(Kn)	0.000000	0.000000	1.000000	0.323398
tensile strength(mpa)	-0.698022	-0.079009	0.323398	1.000000

```
<matplotlib.axes._subplots.AxesSubplot at 0x192f8b56cf8>
```

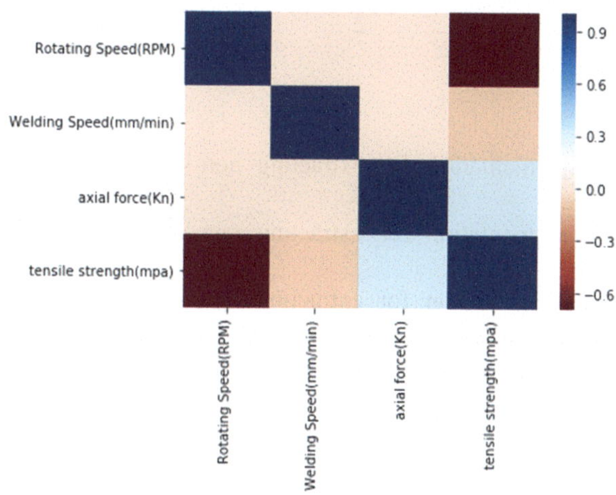

Fig. 3.26 FSW_TS co-relation matrix

	Welding Current (A)	Welding Voltage (V)	Experimental Heat generated (Qe) (J/mm)
Welding Current (A)	1.000000	0.959024	-0.726754
Welding Voltage (V)	0.959024	1.000000	-0.787903
Experimental Heat generated (Qe) (J/mm)	-0.726754	-0.787903	1.000000

```
<matplotlib.axes._subplots.AxesSubplot at 0x14784f23b38>
```

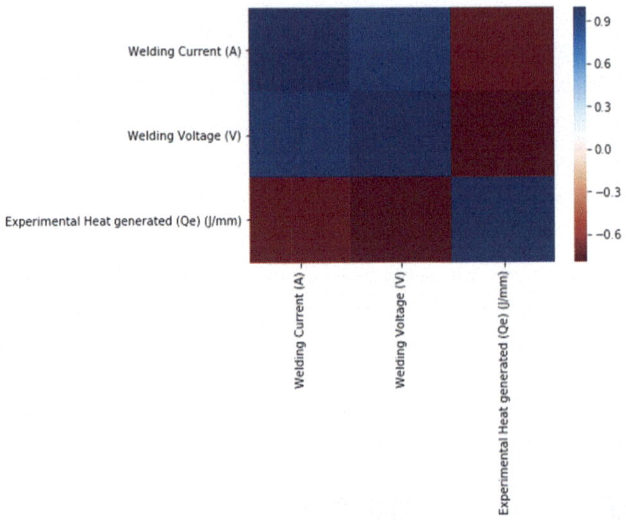

Fig. 3.27 FSW_EHG co-relation matrix

Figure 3.34 shows the implementation for extracting non-linear features of tensile strength.

Figure 3.35 shows the implementation for extracting non-linear features of Experimental Heat Generated.

Figure 3.36 shows the implementation for extracting non-linear features of Torque Developed.

Figure 3.37 shows the implementation for extracting non-linear features of Theoretical Heat Input.

Figure 3.38 shows the implementation for extracting non-linear features of Elongation.

Figure 3.39 shows the implementation for extracting non-linear features of UTS.

Figure 3.40 shows the implementation for extracting non-linear features of Yield Strength.

Figure 3.41 shows the implementation for extracting non-linear features of Wear.

	Welding Current (A)	Welding Voltage (V)	Torque Developed (Nm/mm)
Welding Current (A)	1.000000	0.997449	0.999006
Welding Voltage (V)	0.997449	1.000000	0.999437
Torque Developed (Nm/mm)	0.999006	0.999437	1.000000

```
<matplotlib.axes._subplots.AxesSubplot at 0x25775b59e48>
```

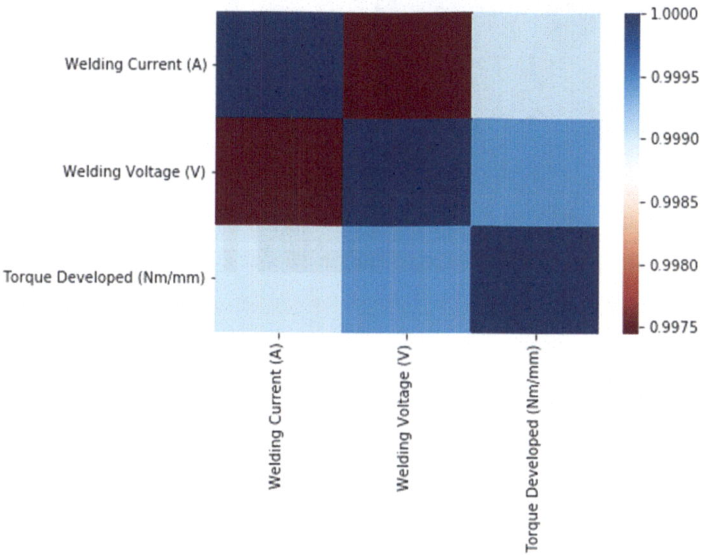

Fig. 3.28 FSW_TD co-relation matrix

3.6.4 *Data Preprocessing Using Scikit-Learn*

In this section, we will look into various linear models for regression and classification using scikit-learn.

The algorithms which will be discussed in the section to give clarity about predictive analysis are

	Welding Current (A)	Welding Voltage (V)	Theoretical Heat Input (Qt) (J/mm)
Welding Current (A)	1.000000	0.959024	-0.755455
Welding Voltage (V)	0.959024	1.000000	-0.814420
Theoretical Heat Input (Qt) (J/mm)	-0.755455	-0.814420	1.000000

```
<matplotlib.axes._subplots.AxesSubplot at 0x225849fca20>
```

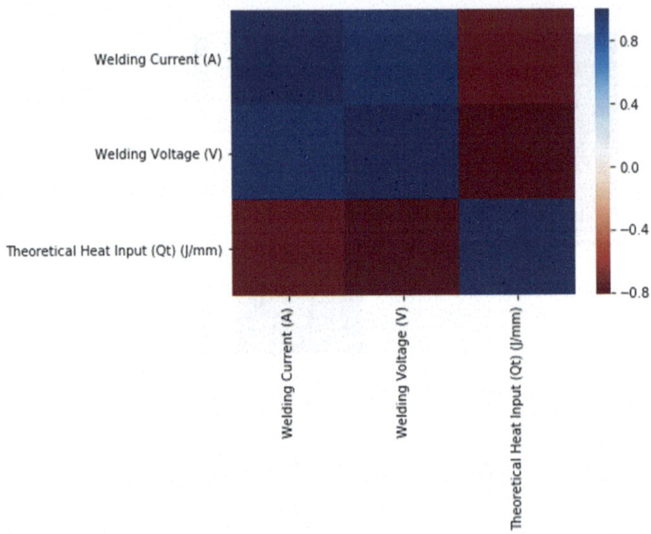

Fig. 3.29 FSW_THI co-relation matrix

Linear Regression

The fundamental concepts of linear regression along with the mathematical formulae based on which this work is performed has been discussed in Chap. 1 (Sect. 1.5).

A standard procedure of splitting the data into 80% of training data and 20% of testing data from the dataset is followed to predict the dependent parameters across each table. It is a two-step process, in which both the steps are a common process to split the dataset into training set and testing set.

	Transverse speed (mm/min)	Tool rotational speed (rpm)	Ɛ%
Transverse speed (mm/min)	1.000000	0.00000	-0.495372
Tool rotational speed (rpm)	0.000000	1.00000	0.860720
Ɛ%	-0.495372	0.86072	1.000000

```
<matplotlib.axes._subplots.AxesSubplot at 0x238d37a7ac8>
```

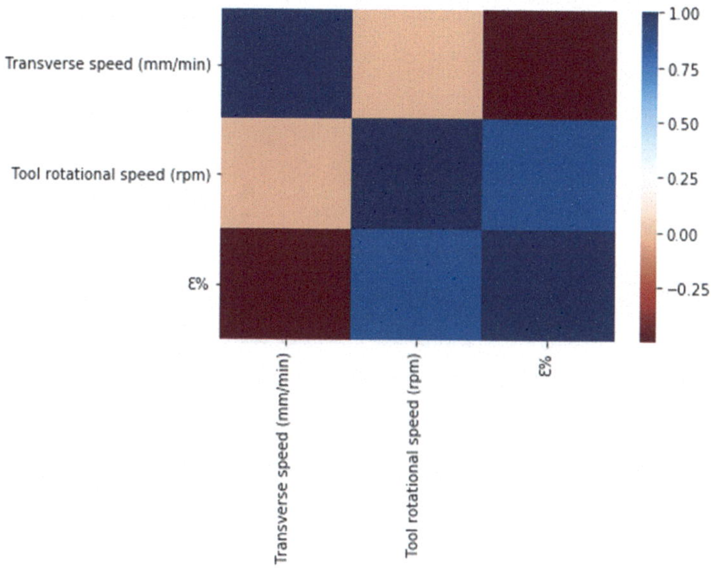

Fig. 3.30 FSW_EL co-relation matrix

Figures 3.42, 3.43, 3.44, 3.45, 3.46, 3.47, 3.48, 3.49, 3.50 show the splitting of each dataset used for training the model.

Figure 3.43 shows that 80% of the data from the FSW_TS dataset will be used for training and the remaining 20% of the data will be used for testing the prediction of the trained model.

	Transverse speed (mm/min)	Tool rotational speed (rpm)	UTS(MPa)
Transverse speed (mm/min)	1.000000	0.000000	0.582879
Tool rotational speed (rpm)	0.000000	1.000000	-0.672095
UTS(MPa)	0.582879	-0.672095	1.000000

```
<matplotlib.axes._subplots.AxesSubplot at 0x1fe67ec6eb8>
```

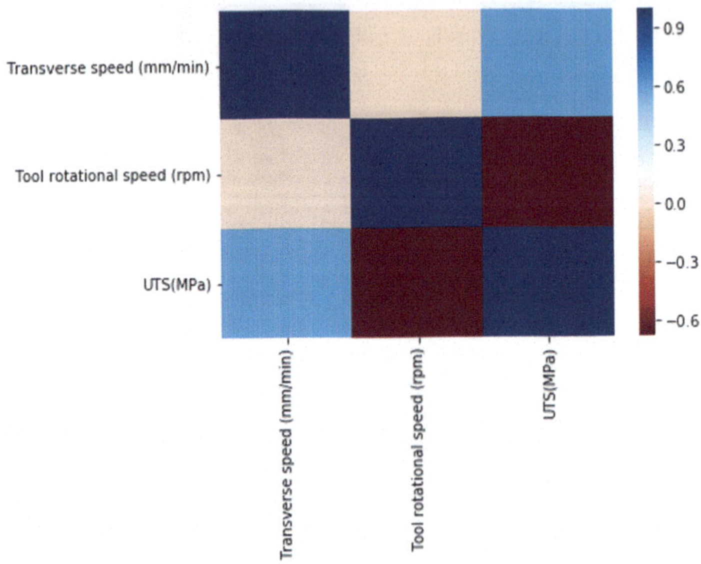

Fig. 3.31 FSW_UTS co-relation matrix

Figure 3.44 shows that 80% of the data from the FSW_EHG Dataset will be used for training and the remaining 20% of the data will be used for testing the prediction of the trained model.

Figure 3.45 shows that 80% of the data from the FSW_TD Dataset will be used for training and the remaining 20% of the data will be used for testing the prediction of the trained model.

	Transverse speed (mm/min)	Tool rotational speed (rpm)	YS(MPa)
Transverse speed (mm/min)	1.000000	0.000000	0.767195
Tool rotational speed (rpm)	0.000000	1.000000	-0.308915
YS(MPa)	0.767195	-0.308915	1.000000

```
<matplotlib.axes._subplots.AxesSubplot at 0x22e5ff35320>
```

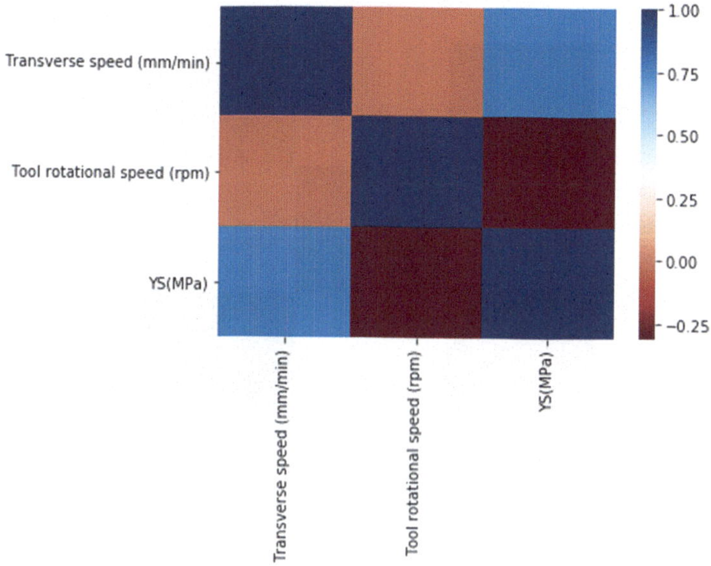

Fig. 3.32 FSW_YS co-relation matrix

Figure 3.46 shows that 80% of the data from the FSW_THI Dataset will be used for training and the remaining 20% of the data will be used for testing the prediction of the trained model.

Figure 3.47 shows that 80% of the data from the FSW_EL Dataset will be used for training and the remaining 20% of the data will be used for testing the prediction of the trained model.

	Time (s)	Frictional Force (N)	Wear (μm)
Time (s)	1.000000	0.941234	0.887643
Frictional Force (N)	0.941234	1.000000	0.964557
Wear (μm)	0.887643	0.964557	1.000000

```
<matplotlib.axes._subplots.AxesSubplot at 0x16c6ecfb828>
```

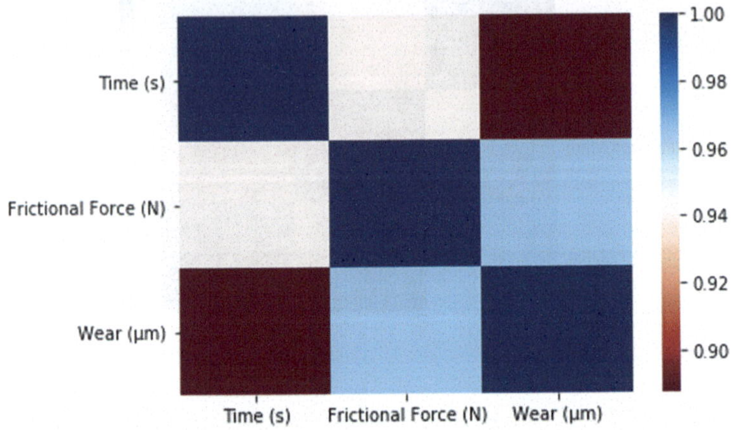

Fig. 3.33 FSW_W co-relation matrix

Figure 3.48 shows that 80% of the data from the FSW_UTS Dataset will be used for training and the remaining 20% of the data will be used for testing the prediction of the trained model.

Figure 3.49 shows that 80% of the data from the FSW_YS Dataset will be used for training and the remaining 20% of the data will be used for testing the prediction of the trained model.

```
# get the summary
desc_df = ts_df.describe()
# add the standard deviation metric
desc_df.loc['+3_std'] = desc_df.loc['mean'] + (desc_df.loc['std'] * 3)
desc_df.loc['-3_std'] = desc_df.loc['mean'] - (desc_df.loc['std'] * 3)
# display it
desc_df
```

	Rotating Speed(RPM)	Welding Speed(mm/min)	axial force(Kn)	tensile strength(mpa)
count	45.000000	45.000000	45.000000	45.000000
mean	1300.000000	50.000000	5.000000	287.971667
std	286.038777	8.257228	1.651446	30.789655
min	900.000000	40.000000	3.000000	217.480000
25%	1100.000000	40.000000	3.000000	267.030000
50%	1300.000000	50.000000	5.000000	285.003333
75%	1500.000000	60.000000	7.000000	307.370000
max	1700.000000	60.000000	7.000000	358.430000
+3_std	2158.116330	74.771685	9.954337	380.340632
-3_std	441.883670	25.228315	0.045663	195.602701

```
# filter the data frame to remove the values exceeding 3 standard deviations
econ_remove_df = ts_df[(np.abs(stats.zscore(ts_df)) < 3).all(axis=1)]
# what rows were removed
ts_df.index.difference(econ_remove_df.index)
```

```
Int64Index([], dtype='int64')
```

Fig. 3.34 Instructions to preprocess the FSW_TS Dataset

Figure 3.50 shows that 80% of the data from the FSW_W Dataset will be used for training and the remaining 20% of the data will be used for testing the prediction of the trained model (Fig. 3.51).

Once the model is trained, we use the trained model to predict the data in the following manner by the help of the following instructions:

```
# get the summary
desc_df = ts_df.describe()
# add the standard deviation metric
desc_df.loc['+3_std'] = desc_df.loc['mean'] + (desc_df.loc['std'] * 3)
desc_df.loc['-3_std'] = desc_df.loc['mean'] - (desc_df.loc['std'] * 3)
# display it
desc_df
```

	Welding Current (A)	Welding Voltage (V)	Experimental Heat generated (Qe) (J/mm)
count	8.000000	8.000000	8.000000
mean	52.250000	167.500000	58.036250
std	1.581139	12.247449	25.433904
min	50.000000	150.000000	31.310000
25%	51.500000	158.750000	35.970000
50%	52.500000	167.500000	55.540000
75%	53.250000	176.250000	77.410000
max	54.000000	185.000000	90.960000
+3_std	56.993416	204.242346	134.337961
-3_std	47.506584	130.757654	-18.265461

```
# filter the data frame to remove the values exceeding 3 standard deviations
econ_remove_df = ts_df[(np.abs(stats.zscore(ts_df)) < 3).all(axis=1)]
# what rows were removed
ts_df.index.difference(econ_remove_df.index)
```

```
C:\Users\LearningBee\Anaconda3\lib\site-packages\ipykernel_launcher.py:2: RuntimeWarning: invalid value encountered in less
```

```
Int64Index([ 0,  1,  2,  3,  4,  5,  6,  7,  8,  9, 10, 11, 12, 13, 14, 15, 16,
            17, 18, 19, 20, 21, 22, 23, 24, 25, 26, 27, 28, 29, 30, 31],
           dtype='int64')
```

Fig. 3.35 Instructions to preprocess the FSW_EHG Dataset

The predictions for each dataset are observed as shown in the following Tables 3.9, 3.10, 3.11, 3.12, 3.13, 3.14, 3.15 and 3.16 mentioned below:

From Fig. 3.52 it is observed that the actual value is quite in agreement with the predicted values of TS.

From Fig. 3.53 it is observed that the actual value is quite in agreement with the predicted values of EHG.

From Fig. 3.54 it is observed that the actual value is quite in agreement with the predicted values of TD.

From Fig. 3.55 it is observed that the actual value is quite in agreement with the predicted values of THI.

```
# get the summary
desc_df = ts_df.describe()
# add the standard deviation metric
desc_df.loc['+3_std'] = desc_df.loc['mean'] + (desc_df.loc['std'] * 3)
desc_df.loc['-3_std'] = desc_df.loc['mean'] - (desc_df.loc['std'] * 3)
# display it
desc_df
```

	Welding Current (A)	Welding Voltage (V)	Torque Developed (Nm/mm)
count	32.000000	32.000000	32.000000
mean	58.437500	227.500000	3.007781
std	4.792047	46.904158	1.096667
min	50.000000	150.000000	1.120000
25%	54.750000	188.750000	2.128522
50%	58.500000	227.500000	3.013729
75%	62.250000	266.250000	3.900307
max	66.000000	305.000000	4.786886
+3_std	72.813640	368.212473	6.297782
-3_std	44.061360	86.787527	-0.282221

```
# filter the data frame to remove the values exceeding 3 standard deviations
econ_remove_df = ts_df[(np.abs(stats.zscore(ts_df)) < 3).all(axis=1)]
# what rows were removed
ts_df.index.difference(econ_remove_df.index)
```

```
Int64Index([], dtype='int64')
```

Fig. 3.36 Instructions to preprocess the FSW_TD Dataset

```
# get the summary
desc_df = ts_df.describe()
# add the standard deviation metric
desc_df.loc['+3_std'] = desc_df.loc['mean'] + (desc_df.loc['std'] * 3)
desc_df.loc['-3_std'] = desc_df.loc['mean'] - (desc_df.loc['std'] * 3)
# display it
desc_df
```

	Welding Current (A)	Welding Voltage (V)	Theoretical Heat Input (Qt) (J/mm)
count	8.000000	8.000000	8.000000
mean	52.250000	167.500000	59.766250
std	1.581139	12.247449	25.142167
min	50.000000	150.000000	33.000000
25%	51.500000	158.750000	38.085000
50%	52.500000	167.500000	57.780000
75%	53.250000	176.250000	80.380000
max	54.000000	185.000000	89.650000
+3_std	56.993416	204.242346	135.192751
-3_std	47.506584	130.757654	-15.660251

```
# filter the data frame to remove the values exceeding 3 standard deviations
econ_remove_df = ts_df[(np.abs(stats.zscore(ts_df)) < 3).all(axis=1)]
# what rows were removed
ts_df.index.difference(econ_remove_df.index)
```

```
Int64Index([], dtype='int64')
```

Fig. 3.37 Instructions to preprocess the FSW_THI Dataset

```
# get the summary
desc_df = ts_df.describe()
# add the standard deviation metric
desc_df.loc['+3_std'] = desc_df.loc['mean'] + (desc_df.loc['std'] * 3)
desc_df.loc['-3_std'] = desc_df.loc['mean'] - (desc_df.loc['std'] * 3)
# display it
desc_df
```

	Transverse speed (mm/min)	Tool rotational speed (rpm)	Ɛ%
count	45.000000	45.000000	45.000000
mean	60.000000	2000.000000	23.493778
std	8.257228	873.862898	4.897829
min	50.000000	600.000000	12.645000
25%	50.000000	1200.000000	19.634444
50%	60.000000	2000.000000	23.493778
75%	70.000000	2800.000000	27.309278
max	70.000000	3400.000000	33.185944
+3_std	84.771685	4621.588693	38.187263
-3_std	35.228315	-621.588693	8.800292

Fig. 3.38 Instructions to preprocess the FSW_EL Dataset

```
# get the summary
desc_df = ts_df.describe()
# add the standard deviation metric
desc_df.loc['+3_std'] = desc_df.loc['mean'] + (desc_df.loc['std'] * 3)
desc_df.loc['-3_std'] = desc_df.loc['mean'] - (desc_df.loc['std'] * 3)
# display it
desc_df
```

	Transverse speed (mm/min)	Tool rotational speed (rpm)	UTS(MPa)
count	36.000000	36.000000	36.000000
mean	60.000000	1700.000000	95.111111
std	8.280787	700.204052	10.418238
min	50.000000	600.000000	76.777778
25%	50.000000	1150.000000	87.777778
50%	60.000000	1700.000000	94.555555
75%	70.000000	2250.000000	100.444444
max	70.000000	2800.000000	124.000000
+3_std	84.842360	3800.612156	126.365825
-3_std	35.157640	-400.612156	63.856397

```
# filter the data frame to remove the values exceeding 3 standard deviations
econ_remove_df = ts_df[(np.abs(stats.zscore(ts_df)) < 3).all(axis=1)]
# what rows were removed
ts_df.index.difference(econ_remove_df.index)
```

```
Int64Index([], dtype='int64')
```

Fig. 3.39 Instructions to preprocess the FSW_UTS Dataset

```
# get the summary
desc_df = ts_df.describe()
# add the standard deviation metric
desc_df.loc['+3_std'] = desc_df.loc['mean'] + (desc_df.loc['std'] * 3)
desc_df.loc['-3_std'] = desc_df.loc['mean'] - (desc_df.loc['std'] * 3)
# display it
desc_df
```

	Transverse speed (mm/min)	Tool rotational speed (rpm)	YS(MPa)
count	36.000000	36.000000	36.000000
mean	60.000000	1700.000000	89.166667
std	8.280787	700.204052	3.777754
min	50.000000	600.000000	82.000000
25%	50.000000	1150.000000	86.083333
50%	60.000000	1700.000000	89.166666
75%	70.000000	2250.000000	92.041667
max	70.000000	2800.000000	96.000000
+3_std	84.842360	3800.612156	100.499930
-3_std	35.157640	-400.612156	77.833403

```
# filter the data frame to remove the values exceeding 3 standard deviations
econ_remove_df = ts_df[(np.abs(stats.zscore(ts_df)) < 3).all(axis=1)]
# what rows were removed
ts_df.index.difference(econ_remove_df.index)
```

```
Int64Index([], dtype='int64')
```

Fig. 3.40 Instructions to preprocess the FSW_YS Dataset

```
# get the summary
desc_df = ts_df.describe()
# add the standard deviation metric
desc_df.loc['+3_std'] = desc_df.loc['mean'] + (desc_df.loc['std'] * 3)
desc_df.loc['-3_std'] = desc_df.loc['mean'] - (desc_df.loc['std'] * 3)
# display it
desc_df
```

	Time (s)	Frictional Force (N)	Wear (μm)
count	21.000000	21.000000	21.000000
mean	40.188810	3.115238	4.774762
std	6.082013	0.061043	0.928324
min	30.386000	3.020000	3.910000
25%	35.288000	3.070000	4.050000
50%	40.189000	3.090000	4.180000
75%	45.089000	3.150000	5.540000
max	49.992000	3.240000	6.420000
+3_std	58.434849	3.298366	7.559735
-3_std	21.942770	2.932111	1.989789

```
# filter the data frame to remove the values exceeding 3 standard deviations
econ_remove_df = ts_df[(np.abs(stats.zscore(ts_df)) < 3).all(axis=1)]
# what rows were removed
ts_df.index.difference(econ_remove_df.index)
```

```
Int64Index([], dtype='int64')
```

Fig. 3.41 Instructions to preprocess the FSW_W Dataset

```
# define our input variable (X) & output variable (Y)
X = ts_df.iloc[:, :-1]
Y = ts_df.iloc[:, 3]
Y

# Split X and y into X_
X_train, X_test, y_train, y_test = train_test_split(X, Y, test_size=.20, random_state=0)
#X_train, X_test, y_train, y_test = model_selection.train_test_split(X, y, train_size=0.75,test_size=0.25, random_state=101)
print ("X_train: ", X_train)
print ("y_train: ", y_train)
print("X_test: ", X_test)
print ("y_test: ", y_test)

# create a Linear Regression model object
regression_model = LinearRegression()

# pass through the X_train & y_train data set
regression_model.fit(X_train, y_train)
```

Fig. 3.42 Instructions to split the dataset into training and test data

```
X_train:        Rotating Speed(RPM)  Welding Speed(mm/min)  axial force(Kn)
31                  1500.0                 50.0                  5.0
29                  1500.0                 40.0                  7.0
20                  1300.0                 40.0                  7.0
41                  1700.0                 50.0                  7.0
2                    900.0                 40.0                  7.0
43                  1700.0                 60.0                  5.0
18                  1300.0                 40.0                  3.0
15                  1100.0                 60.0                  3.0
22                  1300.0                 50.0                  5.0
16                  1100.0                 60.0                  5.0
40                  1700.0                 50.0                  5.0
8                    900.0                 60.0                  7.0
13                  1100.0                 50.0                  5.0
5                    900.0                 50.0                  7.0
17                  1100.0                 60.0                  7.0
32                  1500.0                 50.0                  7.0
14                  1100.0                 50.0                  7.0
35                  1500.0                 60.0                  7.0
7                    900.0                 60.0                  5.0
34                  1500.0                 60.0                  5.0
1                    900.0                 40.0                  5.0
26                  1300.0                 60.0                  7.0
12                  1100.0                 50.0                  3.0
33                  1500.0                 60.0                  3.0
24                  1300.0                 60.0                  3.0
6                    900.0                 60.0                  3.0
23                  1300.0                 50.0                  7.0
36                  1700.0                 40.0                  3.0
21                  1300.0                 50.0                  3.0
19                  1300.0                 40.0                  5.0
9                   1100.0                 40.0                  3.0
39                  1700.0                 50.0                  3.0
42                  1700.0                 60.0                  3.0
3                    900.0                 50.0                  3.0
0                    900.0                 40.0                  3.0
44                  1700.0                 60.0                  7.0
y_train:  31     272.944444
29      287.949444
20      288.480000
41      269.976111
2       308.160000
43      254.971111
18      328.880000
15      270.490000
22      285.520000
16      358.430000
40      257.917222
8       352.350000
13      310.920000
5       289.620000
17      326.790000
32      285.003333
14      345.980000
35      282.057222
7       307.370000
34      269.998333
1       311.020000
26      299.390000
12      300.610000
33      257.939444
24      217.480000
6       306.990000
23      298.680000
36      248.804444
```

Fig. 3.43 Instructions to split FSW_TS dataset into 20% test and 80% training data

```
21     262.020000
19     277.410000
9      319.980000
39     245.858333
42     242.912222
3      299.010000
0      323.090000
44     267.030000
Name: tensile strength(mpa), dtype: float64
X_test:       Rotating Speed(RPM)  Welding Speed(mm/min)  axial force(Kn)
30                 1500.0                   50.0              3.0
37                 1700.0                   40.0              5.0
27                 1500.0                   40.0              3.0
4                   900.0                   50.0              5.0
10                 1100.0                   40.0              5.0
25                 1300.0                   60.0              5.0
28                 1500.0                   40.0              5.0
11                 1100.0                   40.0              7.0
38                 1700.0                   40.0              7.0
y_test:  30    260.885556
37     260.863333
27     263.831667
4      306.160000
10     274.560000
25     275.420000
28     275.890556
11     336.160000
38     272.922222
Name: tensile strength(mpa), dtype: float64

LinearRegression(copy_X=True, fit_intercept=True, n_jobs=None,
         normalize=False)
```

Fig. 3.43 (continued)

```
X_train:     Welding Current (A)  Welding Voltage (V)
1                   50.0                155.0
7                   54.0                185.0
3                   52.0                165.0
0                   50.0                150.0
5                   53.0                175.0
4                   53.0                170.0
y_train:  1    74.28
7    32.85
3    90.96
0    72.12
5    38.96
4    37.01
Name: Experimental Heat generated (Qe) (J/mm), dtype: float64
X_test:     Welding Current (A)  Welding Voltage (V)
6                   54.0                180.0
2                   52.0                160.0
y_test:  6    31.31
2    86.80
Name: Experimental Heat generated (Qe) (J/mm), dtype: float64

LinearRegression(copy_X=True, fit_intercept=True, n_jobs=None,
         normalize=False)
```

Fig. 3.44 Instructions to split FSW_EHG Dataset into 20% test and 80% training data

```
X_train:        Welding Current (A)  Welding Voltage (V)
26                    64.0                280.0
20                    61.0                250.0
13                    57.0                215.0
24                    63.0                270.0
5                     53.0                175.0
17                    59.0                235.0
8                     55.0                190.0
30                    66.0                300.0
25                    63.0                275.0
23                    62.0                265.0
1                     50.0                155.0
31                    66.0                305.0
6                     54.0                180.0
4                     53.0                170.0
18                    60.0                240.0
29                    65.0                295.0
19                    60.0                245.0
9                     55.0                195.0
7                     54.0                185.0
27                    64.0                285.0
3                     52.0                165.0
0                     50.0                150.0
21                    61.0                255.0
15                    58.0                225.0
12                    57.0                210.0
y_train:  26    4.254809
20      3.559216
13      2.700106
24      4.022945
5       1.720000
17      3.163835
8       2.168030
30      4.718538
25      4.091292
23      3.859428
1       1.240000
31      4.786886
6       1.980000
4       1.660000
18      3.327352
29      4.555021
19      3.395699
9       2.236377
7       2.010000
27      4.323157
3       1.560000
```

Fig. 3.45 Instructions to split FSW_TD Dataset into 20% test and 80% training data

```
0      1.120000
21     3.627564
21     3.627564
15     2.931970
12     2.631758
Name: Torque Developed (Nm/mm), dtype: float64
X_test:      Welding Current (A)  Welding Voltage (V)
11                  56.0                205.0
22                  62.0                260.0
10                  56.0                200.0
2                   52.0                160.0
16                  59.0                230.0
14                  58.0                220.0
28                  65.0                290.0
y_test:  11    2.468242
22     3.791081
10     2.399894
2      1.500000
16     3.095487
14     2.863623
28     4.486674
Name: Torque Developed (Nm/mm), dtype: float64
```

Fig. 3.45 (continued)

```
X_train:      Welding Current (A)  Welding Voltage (V)
6                   54.0                180.0
2                   52.0                160.0
1                   50.0                155.0
7                   54.0                185.0
3                   52.0                165.0
0                   50.0                150.0
5                   53.0                175.0
4                   53.0                170.0
y_train:  6    33.00
2    88.60
1    77.64
7    34.35
3    89.65
0    75.00
5    40.56
4    39.33
Name: Theoretical Heat Input (Qt) (J/mm), dtype: float64
X_test:  Empty DataFrame
Columns: [Welding Current (A), Welding Voltage (V)]
Index: []
y_test:  Series([], Name: Theoretical Heat Input (Qt) (J/mm), dtype: float64)

LinearRegression(copy_X=True, fit_intercept=True, n_jobs=None,
         normalize=False)
```

Fig. 3.46 Instructions to split FSW_THI Dataset into 20% test and 80% training data

```
X_train:      Transverse speed (mm/min)  Tool rotational speed (rpm)
31                         60.0                        2600.0
29                         70.0                        2400.0
20                         70.0                        1800.0
41                         70.0                        3200.0
2                          70.0                         600.0
43                         60.0                        3400.0
18                         50.0                        1800.0
15                         50.0                        1600.0
22                         60.0                        2000.0
16                         60.0                        1600.0
40                         60.0                        3200.0
8                          70.0                        1000.0
13                         60.0                        1400.0
5                          70.0                         800.0
17                         70.0                        1600.0
32                         70.0                        2600.0
14                         70.0                        1400.0
35                         70.0                        2800.0
7                          60.0                        1000.0
34                         60.0                        2800.0
1                          60.0                         600.0
26                         70.0                        2200.0
12                         50.0                        1400.0
33                         50.0                        2800.0
24                         50.0                        2200.0
6                          50.0                        1000.0
23                         70.0                        2000.0
36                         50.0                        3000.0
21                         50.0                        2000.0
19                         60.0                        1800.0
9                          50.0                        1200.0
39                         50.0                        3200.0
42                         50.0                        3400.0
3                          50.0                         800.0
0                          50.0                         600.0
44                         70.0                        3400.0
y_train:  31     26.388278
29     22.485111
20     19.590611
41     26.344444
2      12.645000
43     30.247611
18     25.467278
15     24.502444
22     23.493778
16     21.564111
40     29.282778
8      16.558000
13     20.599278
5      16.235000
17     18.625778
32     23.449944
14     17.660944
35     24.414778
7      18.245000
34     27.353111
1      15.334000
26     21.520278
12     23.537611
```

Fig. 3.47 Instructions to split FSW_EL Dataset into 20% test and 80% training data

```
33    30.291444
24    27.396944
6     21.232000
23    20.555444
36    31.256278
21    26.432111
19    22.528944
9     22.572778
39    32.221111
42    33.185944
3     19.569000
0     22.267000
44    27.309278
Name: ε%, dtype: float64
X_test:      Transverse speed (mm/min)  Tool rotational speed (rpm)
30                        50.0                       2600.0
37                        60.0                       3000.0
27                        50.0                       2400.0
4                         60.0                        800.0
10                        60.0                       1200.0
25                        60.0                       2200.0
28                        60.0                       2400.0
11                        70.0                       1200.0
38                        70.0                       3000.0
y_test:  30    29.326611
37    28.317944
27    28.361778
4     17.258000
10    19.634444
25    24.458611
28    25.423444
11    16.696111
38    25.379611
Name: ε%, dtype: float64

LinearRegression(copy_X=True, fit_intercept=True, n_jobs=None,
         normalize=False)
```

Fig. 3.47 (continued)

```
X_train:      Transverse speed (mm/min)  Tool rotational speed (rpm)
11                           70                          1200
29                           70                          2400
27                           50                          2400
35                           70                          2800
33                           50                          2800
28                           60                          2400
32                           70                          2600
8                            70                          1000
13                           60                          1400
5                            70                           800
17                           70                          1600
14                           70                          1400
7                            60                          1000
26                           70                          2200
1                            60                           600
12                           50                          1400
25                           60                          2200
24                           50                          2200
6                            50                          1000
23                           70                          2000
4                            60                           800
18                           50                          1800
21                           50                          2000
19                           60                          1800
9                            50                          1200
34                           60                          2800
3                            50                           800
0                            50                           600
y_train:  11    107.444444
29      95.444444
27      80.777778
35      91.444444
33      76.777778
28      88.111111
32      93.444444
8      124.000000
13      98.111111
5      112.000000
17     103.444444
14     105.444444
7       97.000000
26      97.444444
1       98.000000
12      90.777778
25      90.111111
24      82.777778
6       85.000000
23      99.444444
4      106.000000
18      86.777778
21      84.777778
19      94.111111
9       92.777778
34      84.111111
3       95.000000
0      116.000000
```

Fig. 3.48 Instructions to split FSW_UTS Split Dataset into 20% test and 80% training data

```
Name: UTS(MPa), dtype: float64
X_test:       Transverse speed (mm/min)  Tool rotational speed (rpm)
31                                 60                          2600
20                                 70                          1800
16                                 60                          1600
30                                 50                          2600
22                                 60                          2000
15                                 50                          1600
10                                 60                          1200
2                                  70                           600
y_test:  31     86.111111
20     101.444444
16      96.111111
30      78.777778
22      92.111111
15      88.777778
10     100.111111
2      104.000000
Name: UTS(MPa), dtype: float64

LinearRegression(copy_X=True, fit_intercept=True, n_jobs=None,
         normalize=False)
```

Fig. 3.48 (continued)

```
X_train:       Transverse speed (mm/min)  Tool rotational speed (rpm)
11                                 70.0                        1200.0
29                                 70.0                        2400.0
27                                 50.0                        2400.0
35                                 70.0                        2800.0
33                                 50.0                        2800.0
28                                 60.0                        2400.0
32                                 70.0                        2600.0
8                                  70.0                        1000.0
13                                 60.0                        1400.0
5                                  70.0                         800.0
17                                 70.0                        1600.0
14                                 70.0                        1400.0
7                                  60.0                        1000.0
26                                 70.0                        2200.0
1                                  60.0                         600.0
12                                 50.0                        1400.0
25                                 60.0                        2200.0
24                                 50.0                        2200.0
6                                  50.0                        1000.0
23                                 70.0                        2000.0
4                                  60.0                         800.0
18                                 50.0                        1800.0
21                                 50.0                        2000.0
19                                 60.0                        1800.0
9                                  50.0                        1200.0
34                                 60.0                        2800.0
3                                  50.0                         800.0
0                                  50.0                         600.0
y_train:  11     93.500000
29     91.500000
27     84.500000
35     90.833333
33     83.833333
28     88.000000
32     91.166667
8      95.000000
13     89.666667
5      94.000000
```

Fig. 3.49 Instructions to split FSW_YS Split Dataset into 20% test and 80% training data

```
17     92.833333
14     93.166667
7      96.000000
26     91.833333
1      89.000000
12     86.166667
25     88.333333
24     84.833333
6      82.000000
23     92.166667
4      92.000000
18     85.500000
21     85.166667
19     89.000000
9      86.500000
34     87.333333
3      82.000000
0      95.000000

Name: YS(MPa), dtype: float64
X_test:      Transverse speed (mm/min)  Tool rotational speed (rpm)
31                        60.0                       2600.0
20                        70.0                       1800.0
16                        60.0                       1600.0
30                        50.0                       2600.0
22                        60.0                       2000.0
15                        50.0                       1600.0
10                        60.0                       1200.0
2                         70.0                        600.0
y_test:  31    87.666667
20     92.500000
16     89.333333
30     84.166667
22     88.666667
15     85.833333
10     90.000000
2      91.000000
Name: YS(MPa), dtype: float64

LinearRegression(copy_X=True, fit_intercept=True, n_jobs=None,
         normalize=False)
```

Fig. 3.49 (continued)

From Fig. 3.56 it is observed that the actual value is quite in agreement with the predicted values of EL.

From Fig. 3.57 it is observed that the actual value is quite in agreement with the predicted values of UTS.

From Fig. 3.58 it is observed that the actual value is quite in agreement with the predicted values of UTS.

From Fig. 3.59 it is observed that the actual value is quite in agreement with the predicted values of W.

```
X_train:      Time (s)  Frictional Force (N)
8       38.229               3.07
13      43.130               3.13
20      49.992               3.24
1       31.366               3.06
11      41.169               3.11
10      40.189               3.09
14      44.109               3.14
18      48.030               3.21
6       36.269               3.07
19      49.010               3.22
4       34.307               3.07
2       32.348               3.07
5       35.288               3.07
16      46.070               3.17
9       39.208               3.07
7       37.248               3.07
17      47.050               3.19
3       33.328               3.08
0       30.386               3.02
15      45.089               3.15
12      42.150               3.12
y_train:  8     3.97
13      4.92
20      6.42
1       4.15
11      4.36
10      4.18
14      5.23
18      6.29
6       3.91
19      6.40
4       4.04
2       4.17
5       3.97
16      5.81
9       4.05
7       3.92
17      6.08
3       4.11
0       4.13
15      5.54
12      4.62
Name: Wear (μm), dtype: float64
X_test:  Empty DataFrame
Columns: [Time (s), Frictional Force (N)]
Index: []
y_test:  Series([], Name: Wear (μm), dtype: float64)

LinearRegression(copy_X=True, fit_intercept=True, n_jobs=None,
         normalize=False)
```

Fig. 3.50 Instructions to split FSW_W Split Dataset into 20% test and 80% training data

```
# Get multiple predictions
y_predict = regression_model.predict(X)

# Show the predicions
y_predict[:]

df = pd.DataFrame({'Actual': Y, 'Predicted': y_predict})
df
```

Fig. 3.51 Instructions to predict by using the trained model

Table 3.9 Results of TS_Predicted data

TS	TS_Predicted
323.09	311.694902
311.02	323.082552
308.16	334.470202
299.01	308.488143
306.16	319.875793
289.62	331.263443
306.99	305.281384
307.37	316.669034
352.35	328.056684
319.98	296.224422
274.56	307.612072
336.16	318.999723
300.61	293.017663
310.92	304.405313
345.98	315.792964
270.49	289.810904
358.43	301.198554
326.79	312.586205
328.88	280.753943
277.41	292.141593
288.48	303.529243
262.02	277.547184
285.52	288.934834
298.68	300.322484
217.48	274.340425
275.42	285.728075
299.39	297.115725

(continued)

Table 3.9 (continued)

TS	TS_Predicted
263.831667	265.283463
275.890556	276.671113
287.949444	288.058764
260.885556	262.076704
272.944444	273.464354
285.003333	284.852005
257.939444	258.869945
269.998333	270.257595
282.057222	281.645246
248.804444	249.812984
260.863333	261.200634
272.922222	272.588284
245.858333	246.606225
257.917222	257.993875
269.976111	269.381525
242.912222	243.399466
254.971111	254.787116
267.03	266.174766

Table 3.10 Results of EHG_Predicted data

Experimental heat generated (Q_e) (J/mm)	EHG_Predicted
72.12	73.611818
74.28	76.636364
86.8	78.997273
90.96	81.021818
37.01	40.214545
38.96	41.239091
31.31	34.431818
32.85	33.456364

Table 3.11 Results of TD_Predicted data

Torque developed (Nm/mm)	TD_Predicted
1.12	1.143998
1.24	1.215155
1.5	1.466306
1.56	1.537463
1.66	1.698617
1.72	1.769774
1.98	1.930927
2.01	2.002084
2.16803	2.163238
2.236377	2.234395
2.399894	2.395549
2.468242	2.466706
2.631758	2.62786
2.700106	2.699017
2.863623	2.860171
2.93197	2.931328
3.095487	3.092482
3.163835	3.163639
3.327352	3.324792
3.395699	3.395949
3.559216	3.557103
3.627564	3.62826
3.791081	3.789414
3.859428	3.860571
4.022945	4.021725
4.091292	4.092882
4.254809	4.254036
4.323157	4.325193
4.486674	4.486347
4.555021	4.557504
4.718538	4.718657
4.786886	4.789814

Table 3.12 Results of THI_Predicted data

Theoretical heat input (Q_t) (J/mm)	THI_Predicted
75	82.601674
77.64	77.10375
88.6	80.745657
89.65	81.247733
39.33	32.819725
40.56	33.321801
33	34.893792
34.35	32.395869

Table 3.13 Results of Ɛ% _Predicted data

Ɛ%	Ɛ%_Prediction
22.267	19.715897
15.334	16.778139
12.645	13.840382
19.569	20.676987
17.258	17.73923
16.235	14.801472
21.232	21.638078
18.245	18.70032
16.558	15.762563
22.572778	22.599168
19.634444	19.661411
16.696111	16.723653
23.537611	23.560258
20.599278	20.622501
17.660944	17.684744
24.502444	24.521349
21.564111	21.583591
18.625778	18.645834
25.467278	25.482439
22.528944	22.544682
19.590611	19.606924
26.432111	26.44353
23.493778	23.505772
20.555444	20.568015
27.396944	27.40462
24.458611	24.466863
21.520278	21.529105
28.361778	28.365711
25.423444	25.427953

(continued)

Table 3.13 (continued)

Ɛ%	Ɛ%_Prediction
22.485111	22.490196
29.326611	29.326801
26.388278	26.389044
23.449944	23.451286
30.291444	30.287892
27.353111	27.350134
24.414778	24.412377
31.256278	31.248982
28.317944	28.311225
25.379611	25.373467
32.221111	32.210072
29.282778	29.272315
26.344444	26.334557
33.185944	33.171163
30.247611	30.233405
27.309278	27.295648

Table 3.14 Results of UTS_Predicted data

UTS (MPa)	UTS_Predicted
116	99.444492
98	107.322505
104	115.200518
95	97.283475
106	105.161488
112	113.039501
85	95.122458
97	103.000471
124	110.878484
92.777778	92.961441
100.111111	100.839454
107.444444	108.717467
90.777778	90.800424
98.111111	98.678437
105.444444	106.55645
88.777778	88.639407
96.111111	96.51742
103.444444	104.395433
86.777778	86.47839

(continued)

Table 3.14 (continued)

UTS (MPa)	UTS_Predicted
94.111111	94.356403
101.444444	102.234416
84.777778	84.317373
92.111111	92.195386
99.444444	100.073399
82.777778	82.156356
90.111111	90.034369
97.444444	97.912382
80.777778	79.995339
88.111111	87.873352
95.444444	95.751365
78.777778	77.834321
86.111111	85.712334
93.444444	93.590347
76.777778	75.673304
84.111111	83.551317
91.444444	91.42933

Table 3.15 Results of YS_Predicted data

YS (MPa)	YS_Predicted
95	87.747077
89	91.448929
91	95.150781
82	87.354072
92	91.055924
94	94.757776
82	86.961068
96	90.66292
95	94.364772
86.5	86.568063
90	90.269915
93.5	93.971767
86.166667	86.175059
89.666667	89.876911
93.166667	93.578763
85.833333	85.782055
89.333333	89.483907
92.833333	93.185759
85.5	85.38905
89	89.090902
92.5	92.792754

(continued)

Table 3.15 (continued)

YS (MPa)	YS_Predicted
85.166667	84.996046
88.666667	88.697898
92.166667	92.39975
84.833333	84.603041
88.333333	88.304893
91.833333	92.006745
84.5	84.210037
88	87.911889
91.5	91.613741
84.166667	83.817033
87.666667	87.518885
91.166667	91.220737
83.833333	83.424028
87.333333	87.12588
90.833333	90.827732

Table 3.16 Results of W_Predicted data

Wear (μm)	W_Predicted
4.13	3.401317
4.15	4.063082
4.17	4.208574
4.11	4.35412
4.04	4.155547
3.97	4.128993
3.91	4.102439
3.92	4.075939
3.97	4.049385
4.05	4.022885
4.18	4.340477
4.36	4.658096
4.62	4.803615
4.92	4.949161
5.23	5.094734
5.54	5.24028
5.81	5.557872
6.08	5.875491
6.29	6.19311
6.4	6.338657
6.42	6.656221

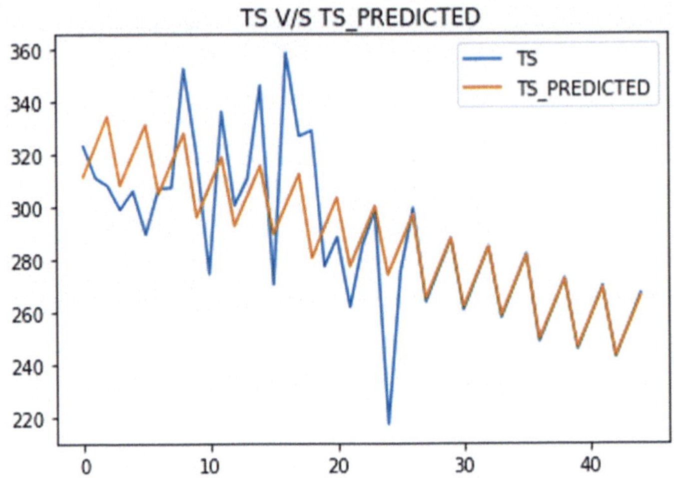

Fig. 3.52 TS versus TS_Predicted

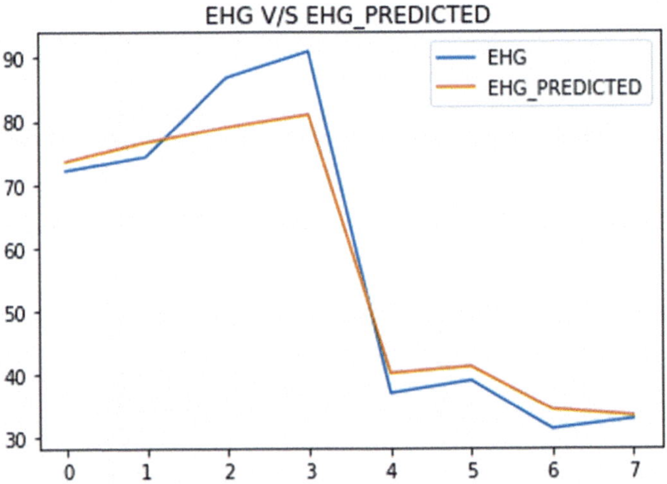

Fig. 3.53 EHG versus EHG_Predicted

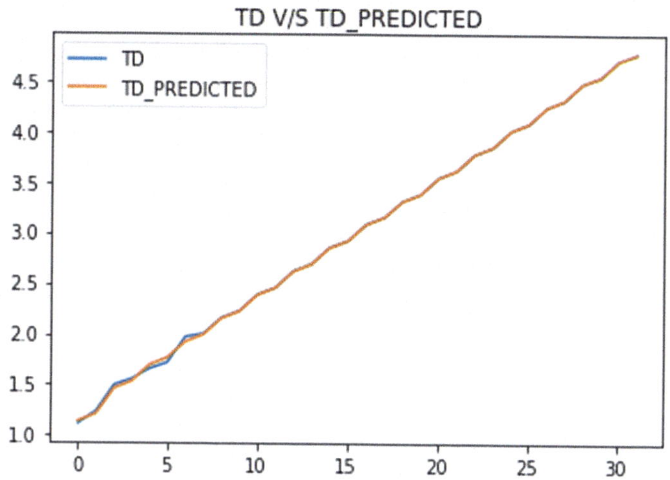

Fig. 3.54 TD versus TD_Predicted

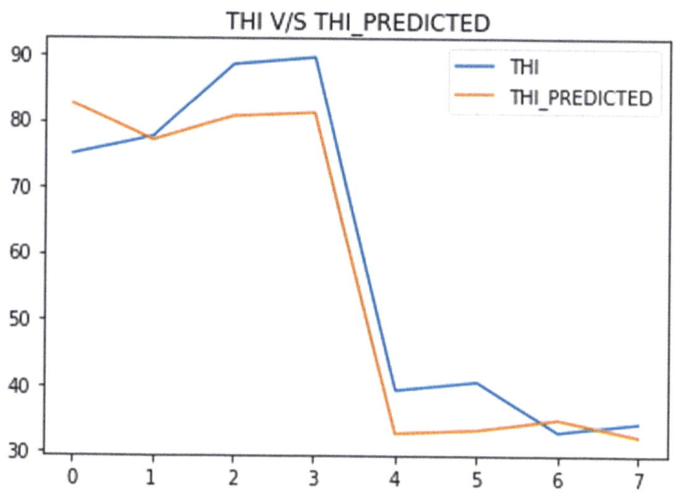

Fig. 3.55 THI versus THI_Predicted

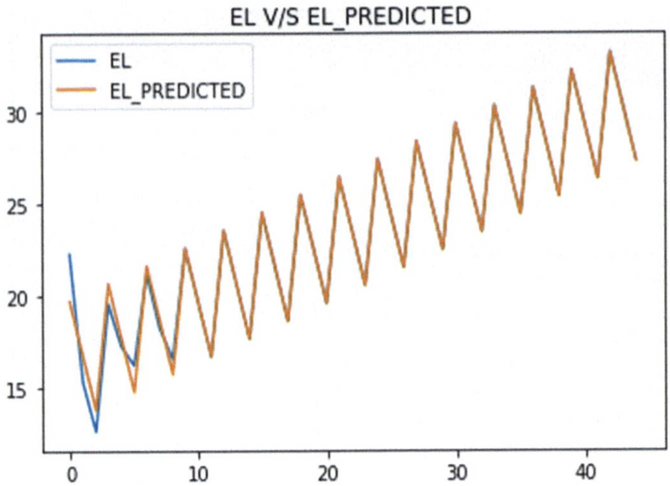

Fig. 3.56 EL versus EL_Predicted

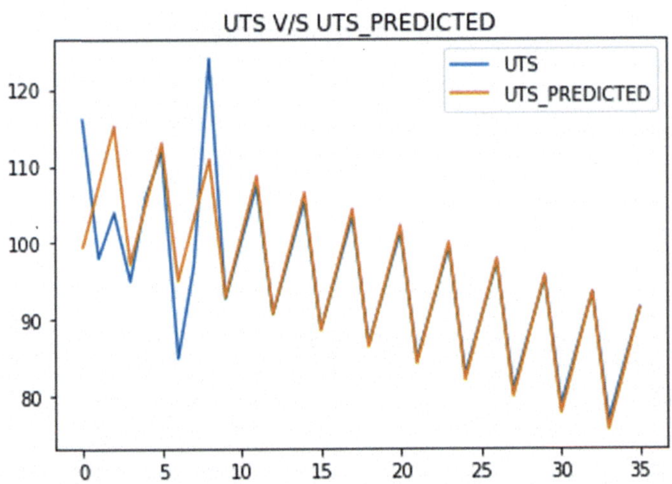

Fig. 3.57 UTS versus UTS_Predicted

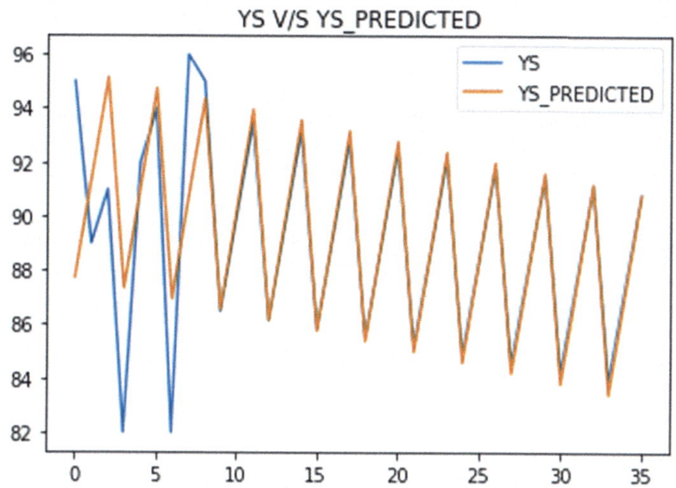

Fig. 3.58 YS versus YS_Predicted

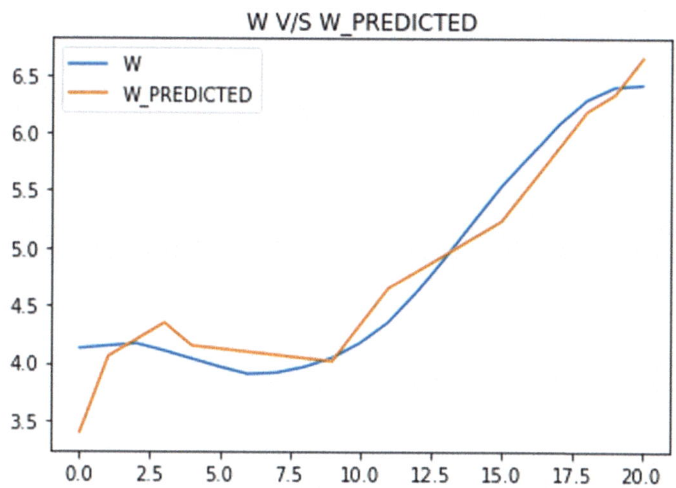

Fig. 3.59 W versus W_Predicted

References

1. Thomas, WM, Nicholas, ED, Needham, JC, Murch, MG, Templesmith, P & Dawes, CJ (1991) Friction stir butt welding GB. Patent Application No. 9125978, pp 42–50
2. Uzun Huseyin (2007) Friction stir welding of SiC particulate reinforced AA2124 aluminium alloy matrix composite. Mater Des 28(5):1440–1446
3. Kalaiselvan K, Dhinaran L, Murugan N (2014) Characterization of friction stir welded boron carbide particulate reinforced AA6061 aluminium alloy stir cast composite. Mater Des 55:176–182
4. Fehrenbacher A, Smith CB, Duffie NA, Ferrier NJ, Pfefferkorn FE, Zinn MR (2014) Combined temperature and force control for robotic friction stir welding. J Manuf Sci Eng 136(2):021–007
5. Mishra RS, Ma ZY (2005) Friction stir welding and processing. Mater Sci Eng R Rep 50(1–2):1–78
6. Lee JA, Carter RW, Ding J (1999) Friction stir welding for aluminum metal matrix composites, NASA report No: NAS 1.15:209876; NASA/TM-1999-209876, p 66
7. Ceschini L, Boromei I, Minak G, Morri A, Tarterini F (2007) Microstructure, tensile and fatigue properties of AA6061/20 vol.% Al_2O_{3p} friction stir welded joints. Compos A Appl Sci Manuf 38(4):1200–1210
8. Fernandez GJ, Murr LE (2004) Characterization of tool wear and weld optimization in the friction-stir welding of cast aluminium 359 + 20% SiC MMC. Mater Charact 52:65–75
9. Xu S, Deng X (2008) A study of texture patterns in friction stir welds. Acta Mater 56 (6):1326–1341
10. Storjohann D, Barabash OM, Babu SS, David SA (2005) Fusion and friction stir welding of aluminum metal matrix composites. Metall Mater Trans A 36(11):3237–3247
11. Rhodes CG, Mahoney MW, Bingel WH, Spurling RA, Bampton CC (1997) Effects of friction stir welding on microstructure of 7075 aluminum. Scripta Mater 36(1):69–75
12. Liu G, Murr LE, Niou CS, McClure JC, Vega FR (1997) Microstructural aspects of the friction-stir welding of 6061-T6 aluminium. Scripta Mater 37(3):335–361
13. Mahoney, MW, Harrigan, WH & Wert, JA (1998a) In: Proceedings of the INALCO'98, Cambridge, UK, vol 2, pp 231–236
14. Liu CH (2000) Structure and properties of boron carbide with aluminum incorporation. Mater Sci Eng, B 72(1):23–26
15. Nandan R, DebRoy T, Bhadeshia HKDH (2008) Recent advances in friction-stir welding—Process, weldment structure and properties. Prog Mater Sci 53(6):980–1023
16. Mahoney MW, Rhodes CG, Flintoff JG, Spurling RA, Bingel WH (1998) Properties of friction-stir-welded 7075 T651 aluminum. Metall Mater Trans A 29(7):1955–1964
17. Prado RA, Murr LE, Shindo DJ, Sota KF (2001) Tool wear in the friction-stir welding of aluminum alloy 6061 + 20 Al_2O_3: a preliminary study. Scripta Mater 45:75
18. Nakata K, Inoki S, Nagano Y, Ushio M (2003) Friction stir welding of Al_2O_3 particulate 6061 Al alloy composite. Mater Sci Forum 426–432:2873–2878
19. Nelson, TW, Zhang, H & Haynes, T (2000) In: Proceedings of the second symposium on friction stir welding, Gothenburg, Sweden, p 42
20. Murrr LE, Li Y, Trillo E, McClure JC (2000) Fundamental issues and industrial applications of friction-stir welding. Mater Technol 15(1):37–48
21. Sharma, SR, Mishra, RS, Mahoney, MW, Jata KV (2001) Friction stir welding and processing, TMS, pp 151–157
22. Palanivel R, Koshy Mathews P, Murugan N, Dinaharan I (2012) Pridection and optimization of wear resistance of friction stir welded dissimilar aluminum alloy. Proc Eng 38:578–584
23. Kim YG, Fujii H, Tsumura T, Komazaki T, Nakata K (2006) Three defects types in FSW of aluminium die casting alloy. Mater Sci Eng A 415(1–2):250–254

24. Nami H, Adgi H, Sharifitabar M, Shamabadi H (2011) Microstructure and mechanical properties of friction stir welded Al/Mg_2Si metal matrix cast composite. Mater Des 32 (2):976–983
25. Elangovan K, Balasubramanian V, Babu S (2009) Predicting tensile strength of friction stir welded AA6061 aluminium alloy joints by a mathematical model. Mater Des 30(1):188–193
26. Cavaliere P, De Santis A, Panella F, Squillace A (2009) Effect of welding parameters on mechanical and microstructural properties of dissimilar AA6082–AA2024 joints produced by friction stir welding. Mater Design 30(3):609–616
27. Mathers G (2002) The welding of aluminium and its alloys. Woodhead Publishing
28. Suban AA, Perumal M, Ayyanar A, Subbiah AV (2017) Microstructural analysis of B 4 C and SiC reinforced Al alloy metal matrix composite joints. Int J Adv Manuf Technol 93(1–4):515–525
29. Xu WF, Liu JH, Chen DL, Luan GH, Yao JS (2012) Improvements of strength and ductility in aluminum alloy joints via rapid cooling during friction stir welding. Mat Sci Eng A 548:89–98
30. Kalaiselvan K, Murugan N (2013) Role of friction stir welding parameters on tensile strength of AA6061–B4C composite joints. Trans Nonferrous Metals Soc China 23(3):616–624
31. Feng AH, Xiao BL, Ma ZY (2008) Effect of microstructural evolution on mechanical properties of friction stir welded AA2009/SiCp composite. Compos Sci Technol 68(9):2141–2148
32. Li JQ, Liu HJ (2013) Effects of welding speed on microstructures and mechanical properties of AA2219-T6 welded by the reverse dual-rotation friction stir welding. Int J Adv Manuf Technol 68(9–12):2071–2083
33. Khodaverdizadeh H, Mahmoudi A, Heidarzadeh A, Nazari E (2012) Effect of friction stir welding (FSW) parameters on strain hardening behavior of pure copper joints. Mater Des 35:330–334
34. Khorrami MS, Kazeminezhad M, Kokabi AH (2012) Microstructure evolutions after friction stir welding of severely deformed aluminum sheets. Mater Des 40:364–372
35. Khorrami MS, Kazeminezhad M, Kokabi AH (2012) Mechanical properties of severely plastic deformed aluminum sheets joined by friction stir welding. Mat Sci Eng A 543:243–248

Chapter 4
Supervised Machine Learning in Wire Cut Electric Discharge Machining (WEDM)

4.1 Introduction

Wire electrical discharge machining (WEDM) technology has grown tremendously since it was first introduced about 40 years back. In 1974, D. H. Dulebohn applied the optical-line follower system to automatically control the shape of the components to be machined by the WEDM process. By 1975, its popularity rapidly increased, as the process and its capabilities were better understood by the industry. It was only towards the end of the 1970s, when computer numerical control (CNC) system was initiated into WEDM, which brought about a major evolution of the machining process [1]. Its broad capabilities have allowed it to encompass the production, aerospace and automotive industries and virtually all areas of conductive material machining. This is because WEDM provides the best alternative or sometimes the only alternative for machining conductive, exotic, high strength and temperature resistive materials, conductive engineering ceramics with the scope of generating intricate shapes and profiles [2, 3].

WEDM has tremendous potential in its applicability in the present-day metal cutting industry for achieving a considerable dimensional accuracy, surface finish and contour generation features of products or parts. Moreover, the cost of wire contributes only 10% of operating cost of WEDM process. The difficulties encountered in the die-sinking EDM are avoided by WEDM, because complex design tool is replaced by moving conductive wire and relative movement of wire guides.

S. A. Vendan et al., *Welding and Cutting Case Studies with Supervised Machine Learning*, Engineering Applications of Computational Methods 1,
https://doi.org/10.1007/978-981-13-9382-2_4

4.2 Process Principle

The mechanism of metal removal in wire electrical discharge machining mainly involves the removal of material due to melting and vaporization caused by the electric spark discharge generated by a pulsating direct current power supply between the electrodes. In WEDM, negative electrode is a continuously moving wire and the positive electrode is the workpiece. The sparks will generate between two closely spaced electrodes under the influence of dielectric liquid. Water is used as dielectric in WEDM, because of its low viscosity and rapid cooling rate [3].

No conclusive theory has been established for the complex machining process. However, empirical evidence suggests that the applied voltage creates an ionized channel between the nearest points of the workpiece and the wire electrodes in the initial stage. In the next stage, the actual discharge takes place with heavy flow of current and the resistance of the ionized channel gradually decreases. The high intensity of current continues to further ionize the channel and a powerful magnetic field is generated. This magnetic field compresses the ionized channel and results in localized heating. Even with sparks of very short duration, the temperature of electrodes can locally rise to very high value which is more than the melting point of the work material due to transformation of the kinetic energy of electrons into heat. The high energy density erodes a part of material from both the wire and workpiece by locally melting and vaporizing and thus it is the dominant thermal erosion process. Figure 4.1 exhibits the schematic diagram of the basic principle of WEDM process [4].

The complete block diagram of WEDM is shown in Fig. 4.2 and the WEDM cutting gap is shown in Fig. 4.3 [5].

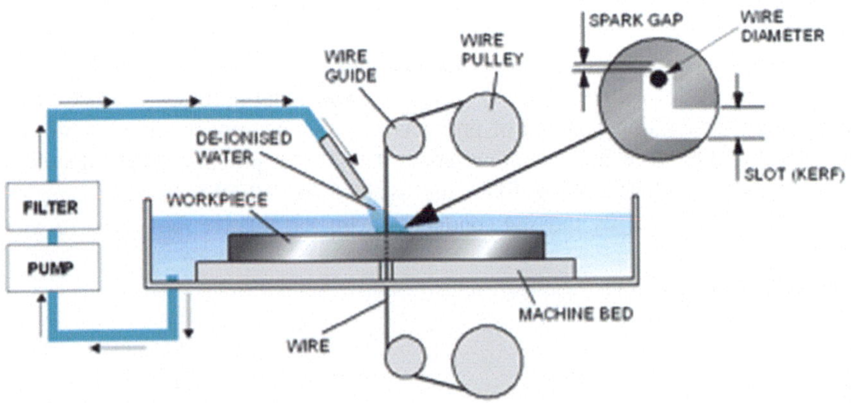

Fig. 4.1 Schematic diagram of the basic principle of WEDM process [4]

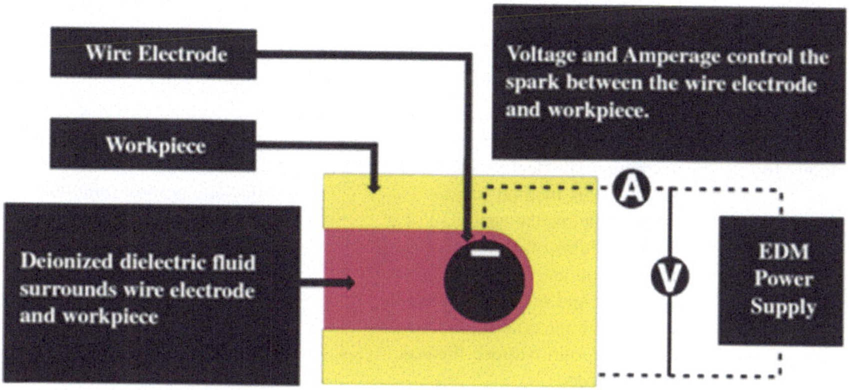

Fig. 4.2 Block diagram of WEDM process [5]

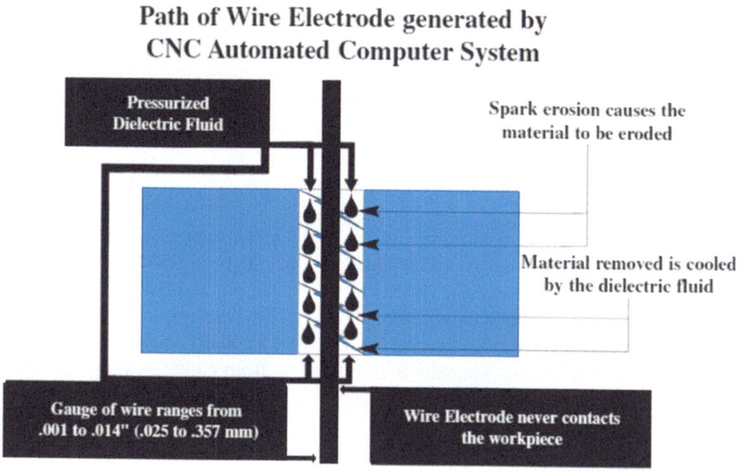

Fig. 4.3 WEDM cutting gap [5]

4.3 Process-State of Art

Important literature discussing the various research attempts made to understand the WEDM process, Parametric influences, Testing and characterization etc., which are briefed in this section (Table 4.1).

Table 4.1 Literature review of WEDM process

Authors and year	Experimentation	Observations
Harshdeep and Ishu Monga [6]	• Taguchi LI6 Orthogonal Array for developing robust design while machining Hll steel was developed • Through this method various process parameters such as Pulse on time, Pulse off time, wire feed, wire tension and Peak current were optimized and output parameters, namely, the material removal rate, wire wear ratio, surface flatness were recorded	• Optimized parametric values facilitated the accomplishment of precise machining • Slight deviations were minimized
Nagaraja et al. [7]	• Investigated the optimization of parameters in WEDM of bronze–alumina MMC • The input parameters considered in this experimental study were pulse on time (TON), pulse off time (TQFF) and wire feed rate • Signal to noise ratio (S/N) and analysis of variance (ANOVA) were used to analyse the effect of the parameters on surface roughness	• The results indicated that the wire feed rate (39.4%) is the most dominant factor affecting the surface roughness
Rao and Venkaiah [8]	• Used central composite face-centred design of response surface methodology (RSM) for their experimentation while optimizing various process parameters on Nimonic-263 alloy • The input parameters taken were pulse on time, pulse off time, peak current and servo voltage • The significance of process parameters is estimated by ANOVA technique • They also developed models for predicting the values of MRR and surface roughness	• The optimal values for MRR and SR from RSM were 3.59856 mm Vmin and 0.363162|xm, respectively • Similarly, the optimal values found from particle swarm optimization (PSO) algorithm for MRR and SR were 3.6713 in Vmin and 0.261 Sum • PSO algorithm delivers more precise results than RSM
Dewangan et al. [9]	• Machined AISI P20 tool steel on WEDM and investigated the effect of various input parameters on surface integrity of the material • The input parameters selected were discharge current (IP), pulse-on time (TON), tool-work time (Tw) and tool-lift time (Tup) • Grey relational analysis (GRA) and fuzzy logic were used to evaluate Grey Fuzzy Reasoning Grade (GFRG)	• Pulse on time followed by discharge current was mostly influencing the results of surface integrity

(continued)

Table 4.1 (continued)

Authors and year	Experimentation	Observations
Azhiri et al. [10]	• Investigated the effect of machining Al/SiC metal matrix composite using dry WEDM process • During the experimentation, the liquid dielectric is replaced with a gaseous medium to enhance the quality of machining environment • Oxygen gas and brass wire were selected as gaseous medium and tool electrode • The effect of the pulse on time, pulse off time, gap voltage, discharge current, wire tension and wire feed were studied on cutting velocity(CV) and surface roughness(SR) using Taguchi's orthogonal array • The relationship between process inputs and responses was correlated using adaptive neuro-fuzzy inference system • Finally, grey relational analysis had been used to optimize CV and SR simultaneously	• Pulse on time and discharge current were found to have a significant effect on CV and SR
Rao et al. [11]	• Used Taguchi method and hybrid genetic algorithm with the aid of linear regression models to optimize surface roughness and material removal rate while machining Aluminium 2014T6 alloy	• Peak current, pulse on time and spark gap majorly influence both SR and MRR • Layer thickness measurement of specimens was done which showed quite high values
Saedon et al. [12]	• Experimented to determine the effects of pulse on time, pulse off time, wire feed and wire tension on the surface roughness (Ra), cutting rate and material removal rate (MRR) in machining of titanium alloy • Combined approach of the orthogonal array and grey relational analysis (GRA) was used for multi-objective optimization of response variables	• The optimal machining settings were found to be as follows: pulse-off time 3p,s, peak current 12A, wire tension 16 N and wire feed 4 mm/min
Goswami and Kumar [13]	• While investigating on Nimonic-80A used Taguchi's methodology to design the experiments to optimize the parameters	• Thick recast layer was formed while using higher pulse on time settings • Lower values of wire deposition were observed when low values of

(continued)

Table 4.1 (continued)

Authors and year	Experimentation	Observations
	• Scanned electron microscopy was performed on machined surfaces to investigate their microstructure	pulse on time and high values of pulse off time were used as settings
Baig and Venkaiah [14]	• Machined Hastelloy C276 by WEDM and using Taguchi and grey relational optimized the parameters • MRR and Kerf width were taken as response variables	• Discharge current (IP) was found to be the most significant factor affecting the MRR and kerf width
Sharma et al. [15]	• Investigate the effect of process parameters on cutting speed and dimensional deviations in cutting high-strength low alloy steel (HSLA) • Response surface methodology (RSM) was used to optimize the parameters • ANOVA was used to determine the significant factors	• Experiments showed that pulse-on time was the most prominent factor affecting the cutting speed and dimensional deviation
Nourbakhsh et al. [16]	• Investigated the effects of pulse width, servo reference voltage, pulse current and wire tension. The response was seen on cutting speed, wire rupture and surface during WEDM of titanium alloys • The influence of zinc-coated brass wire was compared with high-speed brass wire	• Cutting speed increases with pulse width, peak current and pulse interval. Pulse width, wire tension and peak current had a direct impact on surface roughness • Zinc-coated brass wire resulted in higher cutting speed and smoother surface finish as compared to high-speed brass wire • SEM photographs proved that uncoated wire produces a surface finish with more cracks and craters • SEM images of wire reveal ruptures and they are affected by the pulse width and pulse interval
Shandilya et al. [17]	• Worked on response surface methodology (RSM) and artificial neural network (ANN)-based mathematical modelling for the average cutting speed of 10% SiCp/6061 Al metal matrix composite (MMC) during wire electric discharge machining (WEDM) • Servo voltage (SV), pulse-on time (TON), pulse-off time (TOFF) and wire feed rate (WF) were chosen as input factors	• The prediction accuracy of ANN model was about three times better than RSM • Voltage was a more significant parameter on average cutting speed than wire feed rate and pulse-off time

(continued)

Table 4.1 (continued)

Authors and year	Experimentation	Observations
	• A back-propagation neural network model was developed with the help of Box–Behnken design (BBD) of experiments	
Shayan et al. [18]	• Central composite rotatable design (CCRD) to design experiments • Empirical models were developed to create relationships between input factors and their responses • Intelligent models were developed based on back-propagation neural network (BPNN) and these models were compared with mathematical models based on the root mean square error (RMSE) and prediction error percent (PEP) • These models developed were integrated with optimization approaches, namely, desirability function and particle swarm optimization	• Results revealed that air at inlet pressure at Ibar leads to higher MRR and lower Surface Roughness and oversize
Kumar and Agarwal [19]	• Optimized the machining parameters for maximum material removal rate and minimum surface finish by multi-objective genetic algorithm while machining high-speed steel (M2, SKH9) • Zinc-coated copper wire was used as a tool • Pulse peak current, pulse on time, pulse off time, wire feed, wire tension and flushing pressure and their interaction on material removal rate and surface finish was observed • The mathematical models were developed between input parameters and responses by using non-linear regression analysis • These mathematical models were then optimized by using multi-objective optimization technique based on non-dominated sorting genetic algorithm-II to obtain a Pareto-optimal solution set	• MRR and surface finish were influenced more by pulse peak current, pulse duration, pulse-off period and wire feed than by flushing pressure and wire tension

(continued)

Table 4.1 (continued)

Authors and year	Experimentation	Observations
Yang et al. [20]	• Analyse variations in metal removal rate MRR, surface roughness Ra and comer deviation CD while machining pure tungsten with WEDM • This research proposes an effective process parameter optimization approach that integrates Taguchi's parameter design method, response surface methodology (RSM), back-propagation neural network (BPNN) and simulated annealing algorithm (SAA) on WEDM processes • The field-emission SEM	• Built-edge layers were present on the finished surface after the WEDM process • With the increase in pulse on time, the MRR was increased due to increase in number of sparks per second thus deteriorating the surface profile rougher surfaces • Also, increasing the wire tension resulted in the decrease of comer deviation

4.4 Experimentation—Phase I

The material used for this study is AISI 4140. The composition of the material is presented in Table 4.2.

Figure 4.4 shows material removed from the workpiece subjected to WEDM and its CNC display in order to analyse the influence of interdisciplinary process parameters on the material removal rate. The raw materials 60 mm dia and 24 mm height was fixed on the CNC Wire-EDM machine and also feed the CAD diagram for flow of wire.

This machine consists of a power supply, a dielectric system, a computer numerical control system (CNC), a wire guiding system, a five-axis servo control system and a water flushing setup. To supply low energy pulses and high frequency, the anti-electrolysis power supply equipped with the commercial wire-EDM machine has been replaced by the DSP-based high-frequency fine-finish power supply during finish machining. Brass wire with a diameter of 0.25 mm and Zinc-coated wire with a diameter of 0.25 mm are used as the wire electrodes.

The process parameters chosen for this investigation are peak current, pulse on time, pulse off time, wire feed rate, wire tension, servo voltage and servo feed setting. Each of these process parameters is set with three levels, levels are taken based on the manufacture data given to the operator with references. In order the

Table 4.2 Chemical composition of AISI 4140 [21]

Elements	C	Cr	Fe	Me	Mo	p	SI	S
Components	0.38–0.43%	0.80–1.10%	96.78–97.77%	0.75–1.0%	0.15–0.25%	≤0.035%	0.15–0.30%	≤0.040%

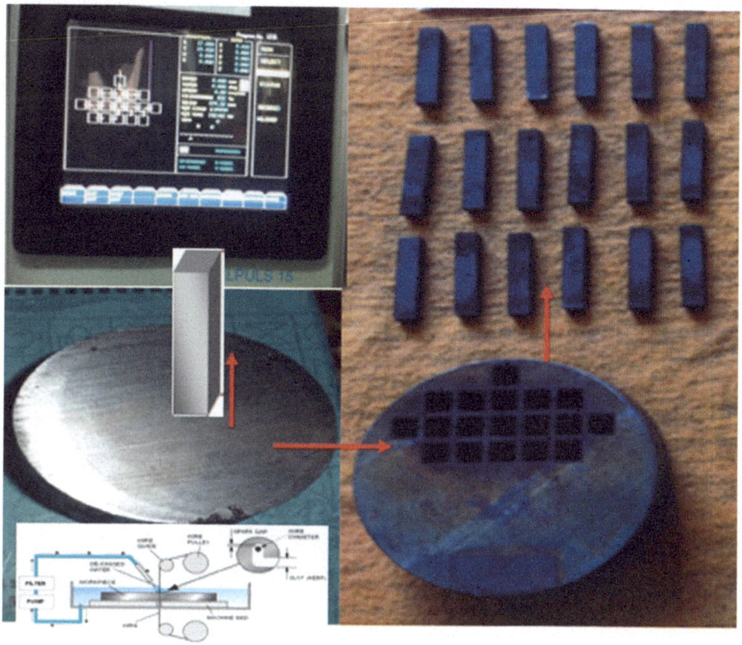

Fig. 4.4 The material removed from the raw workpiece by WEDM and the corresponding CNC display [21]

reduce the number of experiments and the intricacies involved, Taguchi L 18 full factorial experimental design is being implemented. Table 4.3 provides the details of the various trials conducted and the corresponding MRR measured.

Two types of wire Brass and Zinc-coated wire used, the wires are taken 0.25 mm thickness for experimentation. For comparison were prepared in order to investigate the effects of wire-EDM cutting on the fatigue performance of the AZ61 magnesium alloy. The experimental trials conducted along with the parametric values are presented in Table 4.4.

Table 4.3 Process parameters levels [21]

Sl.no	Parameter	Level-1	Level-2	Level-3
1	Peak current (Ip) A	1	2	–
2	Pulse on time (T on) μs	20	23	26
3	Pulse off time (Toff)μs	40	45	50
4	Wire feed rate (Wf) mm/min	2	3	4
5	Wire tension (WT) N	7	8	9
6	Servo voltage (SV) V	15	20	25
7	Servo feed setting (SFS) m/min	95	100	105

Table 4.4 Experimental trials [21]

Ip (A)	Ton (μs)	Toff (μs)	WF (m/min)	WT (N)	SV (V)	SF (mm/min)	MRR (mm^2/min) Brass wire	MRR (mm^2/min) Zin coated wire
1	22	40	5	6	15	2090	0.58	0.77
1	22	45	6	8	20	2100	0.59	0.69
1	22	50	7	10	25	2110	0.53	0.67
1	25	40	5	8	20	2110	0.62	1.02
1	25	45	6	10	25	2090	0.6	0.88
1	25	50	7	6	15	2100	0.64	0.87
1	28	40	6	6	25	2100	0.63	0.85
1	28	45	7	8	15	2110	0.75	0.99
1	28	50	5	10	20	2090	0.69	0.87
2	22	40	7	10	20	2100	0.57	0.76
2	22	45	5	6	25	2110	0.69	0.70
2	22	50	6	8	15	2090	0.59	0.74
2	25	40	6	10	15	2110	0.83	1.01
2	25	45	7	6	20	2090	0.82	0.80
2	25	50	5	8	25	2100	0.59	0.75
2	28	40	7	8	25	2090	0.72	0.87
2	28	45	5	10	15	2100	0.77	0.99
2	28	50	6	6	20	2110	0.69	0.88

4.5 Results and Discussion—Phase I

4.5.1 Metallographic Analysis

The specimens are further subjected to metallographic analysis. SEM image of the specimen (Fig. 4.5) is captured to understand the surface morphology and integrity. It is observed that the molten and oxidized material is flushed away by dielectric, whereas a small amount of molten material is not expelled but rapidly quenched by the dielectric and re-solidifies on the EDM surface to form clustered droplets. The solidified microstructure of the recast layer exhibits droplets and voids in addition to the microcracks.

From the SEM images of the surface characteristics that are showed in Figs. 4.5 and 4.6 reveal high-speed brass wire caused a poor surface finish. Figure 4.6 shows zinc-coated wire produces less ridge on the workpiece surface. Poor surfaces produced by uncoated wire results from the non-uniform thermal load on the samples and produced a recast layer that increases roughness, because of improper cooling and water flushing during the pulse on time, so voids are formed a recast layer. This non-uniform heating is due to non-uniform spark produced and

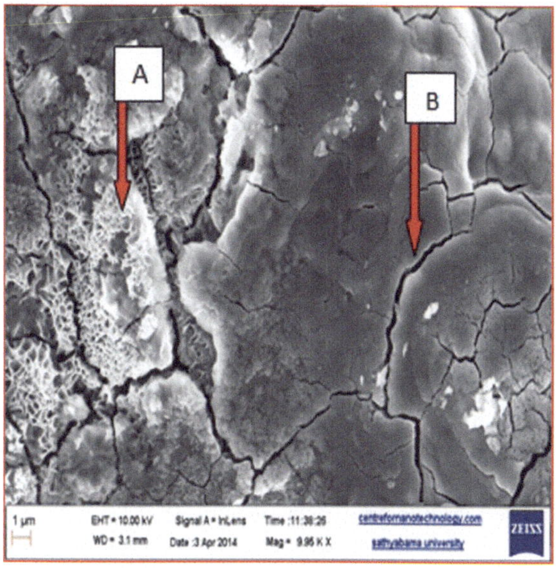

Fig. 4.5 SEM image for microcracks and specimen surface **a**: the recast layer formed during WEDM, **b**: Microcracks on the surface [21]

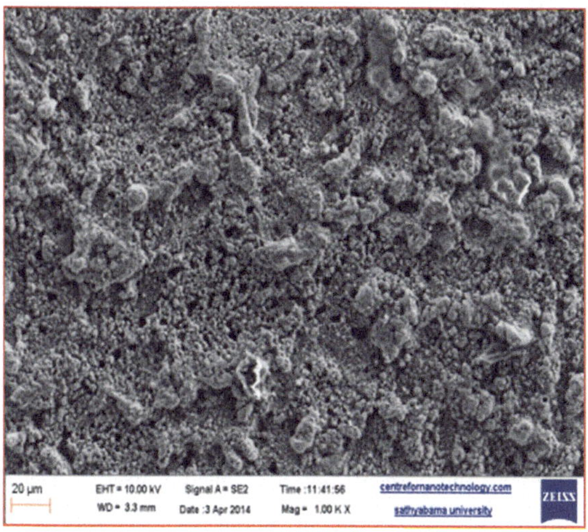

Fig. 4.6 SEM image of first specimen for zinc-coated wire cutting [21]

insufficient flushing rate. Insufficient flushing can also cause arc generation; this can also affect the surface quality, and also more melted drops, globules of debris, craters and cracks. Due to low melting temperature and high heat conductivity, the zinc-coated wire produces better surface quality for Chromoly AISI 4140 materials.

4.5.2 Wire Properties and Influences on Wire-EDM Performance

The wire used in WEDM should provide some features such as high electrical conductivity, sufficient tensile strength at high temperatures, low melting temperature and high heat conductivity in order to reach high performance. The high electrical conductivity helps the wire transfers energy to workpiece efficiently and minimize energy loss of sparking during machining. The wire with high tensile strength is a good heat resistance at high temperature and maintains straight under vibration and tension. The low melting temperature of wire improves the spark formation and decreases dielectric ionization time. The higher thermal conductivity of the electrode ensures a better spark discharge energy distribution during the EDM process. This will increase material removal rate.

Brass wires are an alternative for copper when WEDM users were looking for a better performance. Brass wires are combination of copper and zinc. They are alloyed in different percentages of 35–37% zinc and 63–65% copper. The tensile strength and vapor pressure rating of brass wire increase because of addition of zinc. Also, the percentage of zinc content in brass wire reduces its melting point. The cutting performance and speed improve in machining with brass wire due to stable discharge during machining. During the cutting process, the zinc in the brass wire actually boils off, or vaporizes, which helps cool the wire and delivers more usable energy to the work zone. Therefore, brass wire has become the commonly applied electrode material since its properties progressed in contrast with copper. However, the brass wire conductivity is significantly decreased. Increasing the zinc content in brass wire can enhance the machining speed (more than 40%) but forming a wire with percentage of zinc that is a brittle phase in the alloy will be difficult. In addition, it was disclosed that the gap conductivity will be further enhanced with more addition of zinc that causes drawing/fabrication problems [21].

There are a restricted number of coating materials available for wires in WEDM. Zinc alloy is the most common material that is applied as the wire coated layer. The zinc alloy was chosen because it has a low vaporization temperature compared with the core material. The wire coating is easily vaporized when a pulse is applied and core is protected due to cooling effect of coating material. Also, zinc can only be coated on the core of metallic wire because of low melting point. Hence, brass or copper typically are used as core for these types of wires. An excellent combination of low cost, better flush ability, high mechanical strength and good electrical conductivity are the advantages of zinc-coated wires in contrast to brass or copper wires. In addition, the coated layer decreases the risk of rupturing wires because it protects the core of the wire from thermal shock of electrical discharge. All mentioned advantages for zinc-coated wires enhance the performance of WEDM, increasing in cutting speed and precision.

4.6 Experimentation—Phase II

The material used for this study is Nickel. The composition of the material is presented in Table 4.5.

The process parameters chosen for this investigation are peak current, pulse on time, pulse off time, wire feed rate, water pulsing, wire tension, servo voltage and servo feed setting. Each of these process parameters is set with three levels (Table 4.6).

In order to reduce the number of experiments and the intricacies involved, Taguchi L18 full factorial experimental design is being implemented. Tables 4.7 and 4.8 provide the details of the various trials conducted and the corresponding MRR and surface roughness are measured.

4.7 Results and Discussion—Phase II

The material removed from the raw workpiece by WEDM and the corresponding CNC display is shown in Fig. 4.7.

The specimens are further subjected to metallographic analysis. SEM image of the specimen (Fig. 4.8) is captured to understand the surface morphology and the integrity. It is observed that the molten and oxidized material is flushed away by dielectric, whereas a small amount of molten material is not expelled but rapidly quenched by the dielectric and re-solidifies on the EDM surface to form clustered droplets. The solidified microstructure of the recast layer exhibits droplets and voids in addition to the microcracks.

Table 4.5 Chemical composition of Nickel [22]

	C	Si	Mn	S	Ni	Cu	Fe
Min	–	–	–	–	99.0	–	–
Max	0.02	0.3	0.3	0.01		4.60	0.4

Table 4.6 Process parameters levels [22]

Sl.no	Parameter	Level-1	Level-2	Level-3
1	Peak current (Ip) A	1	2	–
2	Pulse on time (T on) μs	20	23	26
3	Pulse off time (T off) μs	40	45	50
4	Water pressure (Wp) Mps	11	12	13
5	Wire feed rate (Wf) mm/min	2	3	4
6	Wire tension (WT) N	7	8	9
7	Servo voltage (SV) V	15	20	25
8	Servo feed setting (SFS) m/min	95	100	105

Table 4.7 Readings for Brass wire and Nickel material [22]

Ip	Ton	Toff	WP	Wf	Wt	SV	SF	MRR	SR
1	20	40	11	2	7	15	95	1.32	2.2566
1	20	45	12	3	8	20	100	1.09	2.4543
1	20	50	13	4	9	25	105	0.72	1.8824
1	23	40	13	2	8	20	105	1.46	2.2927
1	23	45	11	3	9	25	95	1.12	2.1456
1	23	50	12	4	7	15	100	0.93	2.4008
1	26	40	13	3	7	25	100	1.36	2.3380
1	26	45	11	4	8	15	105	1.29	2.3882
1	26	50	12	2	9	20	95	0.89	2.2094
2	20	40	11	4	9	20	100	1.33	1.9849
2	20	45	12	2	7	25	105	1.2	2.2858
2	20	50	13	3	8	15	95	0.93	2.4696
2	23	40	12	3	9	15	105	1.65	2.4876
2	23	45	13	4	7	20	95	1.3	2.5341
2	23	50	11	2	8	25	100	0.89	2.3896
2	26	40	12	4	8	25	95	1.75	2.6316
2	26	45	13	2	9	15	100	1.82	2.4514
2	26	50	11	3	7	20	105	0.85	2.3337

Table 4.8 Readings for Zinc-coated wire and Nickel material [22]

Ip	Ton	Toff	Wp	Wf	Wt	Sv	Sf	MRR	SR
1	20	40	11	2	7	15	95	1.40	2.1445
1	20	45	12	3	8	20	100	1.2	2.4829
1	20	50	13	4	9	25	105	0.76	2.4520
1	23	40	13	2	8	20	105	1.534	2.5045
1	23	45	11	3	9	25	95	1.263	2.3218
1	23	50	12	4	7	15	100	1.04	2.5483
1	26	40	13	3	7	25	100	1.56	2.5124
1	26	45	11	4	8	15	105	1.59	2.2798
1	26	50	12	2	9	20	95	1.07	2.4732
2	20	40	11	4	9	20	100	1.38	2.2115
2	20	45	12	2	7	25	105	1.05	2.0818
2	20	50	13	3	8	15	95	1.03	2.2525
2	23	40	12	3	9	15	105	1.74	2.4889
2	23	45	13	4	7	20	95	1.40	2.4111
2	23	50	11	2	8	25	100	0.92	2.3660
2	26	40	12	4	8	25	95	1.80	2.4423
2	26	45	13	2	9	15	100	1.68	2.8251
2	26	50	11	3	7	20	105	1.36	2.2223

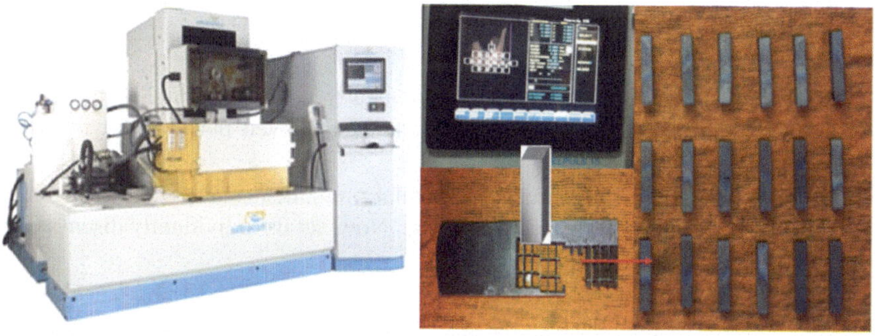

(a) (b)

Fig. 4.7 Material removed from the workpiece subjected to WEDM and its CNC display [22]

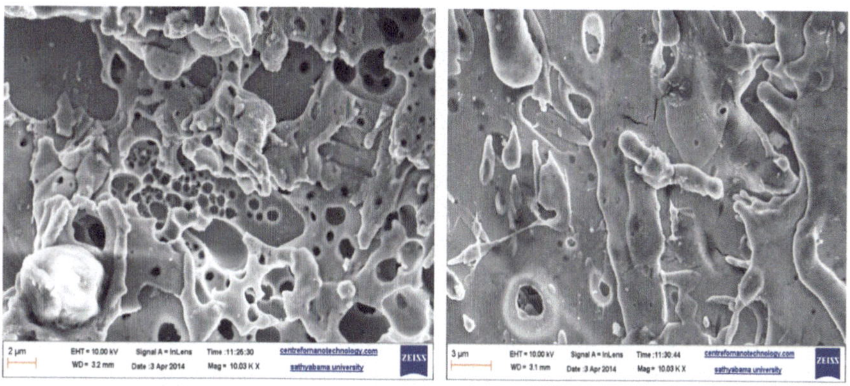

(a) (b)

Fig. 4.8 SEM image for microcracks and specimen surface [22]. **a** Brass wire first Specimen SEM Images, **b** Zinc-coated wire first specimen image

4.8 Parametric Analysis Using Machine Learning Terminologies

This section presents details on FSW data analysis and predictions using machine learning techniques. The fundamental concepts based on which this work is performed has been discussed in Chap. 1 (Sect. 1.5).

4.8.1 Data Analysis Using NumPy *and* Pandas

Since the data structures are going to be frequently used, importing them into the local namespace is equally important and can be achieved in the following ways (Fig. 4.9).

Understanding Pandas requires fundamental prerequisites about its widely used data structures such as Series and DataFrame. Now, let us individually discuss each of these data structures (Fig. 4.10).

The first step is to import the dataset. Figures 4.11, 4.12, 4.13, 4.14, 4.15, 4.16, 4.17, 4.18 and 4.19 give an idea on how to import the dataset using pandas.

4.8.2 Data Visualization Using Seaborn *and* Matplotlib

The analysis and visualization done for the datasets is to find out the co-relation between the independent variables, i.e. inputs and dependent variable, i.e. output by the help of seaborn library (Fig. 4.20). Figures 4.21, 4.22, 4.23, 4.24, 4.25, 4.26, 4.27, 4.28, 4.29 shown are the heat map and co-relation matrix for each dataset.

From the co-relation matrix of MRR mentioned above, the following input parameter such as Ip, Ton and TOW are POSITIVELY co-related to MRR.

From the co-relation matrix of OC mentioned above, the following input parameter such as Ip, Ton and TOW are POSITIVELY co-related to OC.

From the co-relation matrix of SR mentioned above, the following input parameter such as Ip, Ton and TOW are POSITIVELY co-related to SR.

From the co-relation matrix of MRR_B mentioned above, the following input parameter such as Ip, Ton and SF are POSITIVELY co-related to MRR_B.

From the co-relation matrix of MRR_B_N mentioned above, the following input parameter such as Ip, Ton and WP are POSITIVELY co-related to MRR_B_N.

From the co-relation matrix of SR_B_N mentioned above, the following input parameter such as Ip, Ton and WP are POSITIVELY co-related to SR_B_N.

```
import pandas as pd
```

Fig. 4.9 Instructions to import Pandas

```
from pandas import Series, DataFrame
```

Fig. 4.10 Instructions to import series and DataFrame from Pandas

Fig. 4.11 Instructions to import MRR

```
# Load the data and replace the '..' with nan
ts_df = pd.read_excel('MRR.xlsx')

ts_df = ts_df.astype(float)

display(ts_df)
```

	Ip(A)	Ton	TOW	MRR
0	2.0	50.0	70.0	1.5438
1	2.0	50.0	80.0	2.3063
2	2.0	50.0	90.0	1.3125
3	2.0	100.0	70.0	1.7938
4	2.0	100.0	80.0	2.0625
5	2.0	100.0	90.0	2.1500
6	2.0	150.0	70.0	1.9000
7	2.0	150.0	80.0	2.1188
8	2.0	150.0	90.0	2.1250
9	5.0	50.0	70.0	7.3500
10	5.0	50.0	80.0	8.3167
11	5.0	50.0	90.0	8.7968
12	5.0	100.0	70.0	8.1917
13	5.0	100.0	80.0	9.2250
14	5.0	100.0	90.0	9.4250
15	5.0	150.0	70.0	7.6580
16	5.0	150.0	80.0	9.1750
17	5.0	150.0	90.0	9.5080
18	8.0	50.0	70.0	14.2417
19	8.0	50.0	80.0	17.1000
20	8.0	50.0	90.0	17.5167
21	8.0	100.0	70.0	15.8417
22	8.0	100.0	80.0	18.8670
23	8.0	100.0	90.0	20.4670
24	8.0	150.0	70.0	14.5875
25	8.0	150.0	80.0	16.9875
26	8.0	150.0	90.0	19.3500

From the co-relation matrix of MRR_Z mentioned above, the following input parameter such as WT, Ton and SF are POSITIVELY co-related to MRR_Z.

From the co-relation matrix of MRR_Z_N mentioned above, the following input parameter such as Ip, Ton and WF are POSITIVELY co-related to MRR_Z_N.

From the co-relation matrix of SR_Z_N mentioned above, the following input parameter such as Ton and WT are POSITIVELY co-related to SR_Z_N.

Fig. 4.12 Instructions to import OC

```
# Load the data and replace the '..' with nan
ts_df = pd.read_excel('OC.xlsx')

ts_df = ts_df.astype(float)

display(ts_df)
```

	Ip(A)	Ton	TOW	OC
0	2.0	50.0	70.0	0.012
1	2.0	50.0	80.0	0.018
2	2.0	50.0	90.0	0.016
3	2.0	100.0	70.0	0.086
4	2.0	100.0	80.0	0.056
5	2.0	100.0	90.0	0.068
6	2.0	150.0	70.0	0.096
7	2.0	150.0	80.0	0.086
8	2.0	150.0	90.0	0.096
9	5.0	50.0	70.0	0.106
10	5.0	50.0	80.0	0.130
11	5.0	50.0	90.0	0.048
12	5.0	100.0	70.0	0.082
13	5.0	100.0	80.0	0.144
14	5.0	100.0	90.0	0.084
15	5.0	150.0	70.0	0.096
16	5.0	150.0	80.0	0.195
17	5.0	150.0	90.0	0.182
18	8.0	50.0	70.0	0.181
19	8.0	50.0	80.0	0.174
20	8.0	50.0	90.0	0.158
21	8.0	100.0	70.0	0.189
22	8.0	100.0	80.0	0.225
23	8.0	100.0	90.0	0.200
24	8.0	150.0	70.0	0.201
25	8.0	150.0	80.0	0.204
26	8.0	150.0	90.0	0.226

4.8.3 Data Preprocessing Using Scikit-Learn

Data preprocessing is one of the crucial steps in machine learning. This is an unavoidable step incase of data which are in unstructured form. Further to this, we will be discussing data preprocessing by applying machine learning libraries such as scikit-learn.

Fig. 4.13 Instructions to import SR

```
# Load the data and replace the '..' with nan
ts_df = pd.read_excel('SR.xlsx')

ts_df = ts_df.astype(float)

display(ts_df)
```

	Ip(A)	Ton	TOW	SR
0	2.0	50.0	70.0	5.267
1	2.0	50.0	80.0	5.330
2	2.0	50.0	90.0	4.130
3	2.0	100.0	70.0	4.400
4	2.0	100.0	80.0	4.530
5	2.0	100.0	90.0	4.130
6	2.0	150.0	70.0	4.670
7	2.0	150.0	80.0	4.600
8	2.0	150.0	90.0	4.067
9	5.0	50.0	70.0	6.000
10	5.0	50.0	80.0	6.530
11	5.0	50.0	90.0	6.200
12	5.0	100.0	70.0	7.330
13	5.0	100.0	80.0	7.200
14	5.0	100.0	90.0	7.867
15	5.0	150.0	70.0	6.867
16	5.0	150.0	80.0	7.800
17	5.0	150.0	90.0	7.930
18	8.0	50.0	70.0	6.600
19	8.0	50.0	80.0	6.930
20	8.0	50.0	90.0	7.530
21	8.0	100.0	70.0	8.867
22	8.0	100.0	80.0	10.000
23	8.0	100.0	90.0	10.130
24	8.0	150.0	70.0	10.000
25	8.0	150.0	80.0	9.800
26	8.0	150.0	90.0	10.000

This is a preprocessing technique used to derive non-linear features. It is mostly applied along with linear regression algorithm to train the model of higher degree. It is implemented in the following manner.

```
# Load the data and replace the '..' with nan
ts_df = pd.read_excel('MRR_brass_wire.xlsx')

ts_df = ts_df.astype(float)

display(ts_df)
```

	Ip (A)	Ton (μs)	Toff (μs)	WF (m/min)	WT (N)	SV (V)	SF (mm/min)	MRR (mm2/min) Brass wire
0	1.0	22.0	40.0	5.0	6.0	15.0	2090.0	0.58
1	1.0	22.0	45.0	6.0	8.0	20.0	2100.0	0.59
2	1.0	22.0	50.0	7.0	10.0	25.0	2110.0	0.53
3	1.0	25.0	40.0	5.0	8.0	20.0	2110.0	0.62
4	1.0	25.0	45.0	6.0	10.0	25.0	2090.0	0.60
5	1.0	25.0	50.0	7.0	6.0	15.0	2100.0	0.64
6	1.0	28.0	40.0	6.0	6.0	25.0	2100.0	0.63
7	1.0	28.0	45.0	7.0	8.0	15.0	2110.0	0.75
8	1.0	28.0	50.0	5.0	10.0	20.0	2090.0	0.69
9	2.0	22.0	40.0	7.0	10.0	20.0	2100.0	0.57
10	2.0	22.0	45.0	5.0	6.0	25.0	2110.0	0.69
11	2.0	22.0	50.0	6.0	8.0	15.0	2090.0	0.59
12	2.0	25.0	40.0	6.0	10.0	15.0	2110.0	0.83
13	2.0	25.0	45.0	7.0	6.0	20.0	2090.0	0.82
14	2.0	25.0	50.0	5.0	8.0	25.0	2100.0	0.59
15	2.0	28.0	40.0	7.0	8.0	25.0	2090.0	0.72
16	2.0	28.0	45.0	5.0	10.0	15.0	2100.0	0.77
17	2.0	28.0	50.0	6.0	6.0	20.0	2110.0	0.69

Fig. 4.14 Instructions to import MRR_B

Figure 4.30 shows the implementation for extracting non-linear features of MRR.

Figure 4.31 shows the implementation for extracting non-linear features of OC.

Figure 4.32 shows the implementation for extracting non-linear features of OC.

Figure 4.33 shows the implementation for extracting non-linear features of MRR_B.

Figure 4.34 shows the implementation for extracting non-linear features of MRR_B_N.

```
# Load the data and replace the '..' with nan
ts_df = pd.read_excel('MRR_brass_metal_nickel_wire.xlsx')

ts_df = ts_df.astype(float)

display(ts_df)
```

	Ip	Ton	Toff	WP	Wf	Wt	SV	SF	MRR
0	1.0	20.0	40.0	11.0	2.0	7.0	15.0	95.0	1.32
1	1.0	20.0	45.0	12.0	3.0	8.0	20.0	100.0	1.09
2	1.0	20.0	50.0	13.0	4.0	9.0	25.0	105.0	0.72
3	1.0	23.0	40.0	13.0	2.0	8.0	20.0	105.0	1.46
4	1.0	23.0	45.0	11.0	3.0	9.0	25.0	95.0	1.12
5	1.0	23.0	50.0	12.0	4.0	7.0	15.0	100.0	0.93
6	1.0	26.0	40.0	13.0	3.0	7.0	25.0	100.0	1.36
7	1.0	26.0	45.0	11.0	4.0	8.0	15.0	105.0	1.29
8	1.0	26.0	50.0	12.0	2.0	9.0	20.0	95.0	0.89
9	2.0	20.0	40.0	11.0	4.0	9.0	20.0	100.0	1.33
10	2.0	20.0	45.0	12.0	2.0	7.0	25.0	105.0	1.20
11	2.0	20.0	50.0	13.0	3.0	8.0	15.0	95.0	0.93
12	2.0	23.0	40.0	12.0	3.0	9.0	15.0	105.0	1.65
13	2.0	23.0	45.0	13.0	4.0	7.0	20.0	95.0	1.30
14	2.0	23.0	50.0	11.0	2.0	8.0	25.0	100.0	0.89
15	2.0	26.0	40.0	12.0	4.0	8.0	25.0	95.0	1.75
16	2.0	26.0	45.0	13.0	2.0	9.0	15.0	100.0	1.82
17	2.0	26.0	50.0	11.0	3.0	7.0	20.0	105.0	0.85

Fig. 4.15 Instructions to import MRR_B_N

Figure 4.35 shows the implementation for extracting non-linear features of SR_B_N.

Figure 4.36 shows the implementation for extracting non-linear features of MRR_Z.

Figure 4.37 shows the implementation for extracting non-linear features of MRR_Z_N.

Figure 4.38 shows the implementation for extracting non-linear features of SR_Z_N.

```
# Load the data and replace the '..' with nan
ts_df = pd.read_excel('SR_brass_metal_nickel_wire.xlsx')

ts_df = ts_df.astype(float)

display(ts_df)
```

	Ip	Ton	Toff	WP	Wf	Wt	SV	SF	SR
0	1.0	20.0	40.0	11.0	2.0	7.0	15.0	95.0	2.2566
1	1.0	20.0	45.0	12.0	3.0	8.0	20.0	100.0	2.4543
2	1.0	20.0	50.0	13.0	4.0	9.0	25.0	105.0	1.8824
3	1.0	23.0	40.0	13.0	2.0	8.0	20.0	105.0	2.2927
4	1.0	23.0	45.0	11.0	3.0	9.0	25.0	95.0	2.1456
5	1.0	23.0	50.0	12.0	4.0	7.0	15.0	100.0	2.4008
6	1.0	26.0	40.0	13.0	3.0	7.0	25.0	100.0	2.3380
7	1.0	26.0	45.0	11.0	4.0	8.0	15.0	105.0	2.3882
8	1.0	26.0	50.0	12.0	2.0	9.0	20.0	95.0	2.2094
9	2.0	20.0	40.0	11.0	4.0	9.0	20.0	100.0	1.9849
10	2.0	20.0	45.0	12.0	2.0	7.0	25.0	105.0	2.2858
11	2.0	20.0	50.0	13.0	3.0	8.0	15.0	95.0	2.4696
12	2.0	23.0	40.0	12.0	3.0	9.0	15.0	105.0	2.4876
13	2.0	23.0	45.0	13.0	4.0	7.0	20.0	95.0	2.5341
14	2.0	23.0	50.0	11.0	2.0	8.0	25.0	100.0	2.3896
15	2.0	26.0	40.0	12.0	4.0	8.0	25.0	95.0	2.6316
16	2.0	26.0	45.0	13.0	2.0	9.0	15.0	100.0	2.4514
17	2.0	26.0	50.0	11.0	3.0	7.0	20.0	105.0	2.3337

Fig. 4.16 Instructions to import SR_B_N

4.8.4 Prediction Analysis Using Scikit-Learn

In this section, we will look into various linear models for regression and classification using scikit-learn. The algorithms which will be discussed in the section give a clarity about predictive analysis.

Linear Regression

The fundamental concepts of linear regression along with the mathematical formulae based on which this work is performed has been discussed in Chap. 1 (Sect. 1.5).

```
# Load the data and replace the '..' with nan
ts_df = pd.read_excel('MRR_zinc_wire.xlsx')

ts_df = ts_df.astype(float)

display(ts_df)
```

	Ip (A)	Ton (µs)	Toff (µs)	WF (m/min)	WT (N)	SV (V)	SF (mm/min)	MRR (mm2/min)Zin coated wire
0	1.0	22.0	40.0	5.0	6.0	15.0	2090.0	0.77
1	1.0	22.0	45.0	6.0	8.0	20.0	2100.0	0.69
2	1.0	22.0	50.0	7.0	10.0	25.0	2110.0	0.67
3	1.0	25.0	40.0	5.0	8.0	20.0	2110.0	1.02
4	1.0	25.0	45.0	6.0	10.0	25.0	2090.0	0.88
5	1.0	25.0	50.0	7.0	6.0	15.0	2100.0	0.87
6	1.0	28.0	40.0	6.0	6.0	25.0	2100.0	0.85
7	1.0	28.0	45.0	7.0	8.0	15.0	2110.0	0.99
8	1.0	28.0	50.0	5.0	10.0	20.0	2090.0	0.87
9	2.0	22.0	40.0	7.0	10.0	20.0	2100.0	0.76
10	2.0	22.0	45.0	5.0	6.0	25.0	2110.0	0.70
11	2.0	22.0	50.0	6.0	8.0	15.0	2090.0	0.74
12	2.0	25.0	40.0	6.0	10.0	15.0	2110.0	1.01
13	2.0	25.0	45.0	7.0	6.0	20.0	2090.0	0.80
14	2.0	25.0	50.0	5.0	8.0	25.0	2100.0	0.75
15	2.0	28.0	40.0	7.0	8.0	25.0	2090.0	0.87
16	2.0	28.0	45.0	5.0	10.0	15.0	2100.0	0.99
17	2.0	28.0	50.0	6.0	6.0	20.0	2110.0	0.88

Fig. 4.17 Instructions to import MRR_Z

A standard procedure of splitting the data into 80% of training data and 20% of testing data from the dataset is followed to predict the dependent parameters across each table. It is a two-step process, in which both the steps are a common process to split the dataset into training set and testing set (Fig. 4.39).

Figures 4.40, 4.41, 4.42, 4.43, 4.44, 4.45, 4.46, 4.47 and 4.48 show the splitting of each dataset used for training the model.

Figure 4.40 shows that 80% of the data from the MRR Dataset will be used for training and the remaining 20% of the data will be used for testing the prediction of the trained model.

Figure 4.41 shows that 80% of the data from the OC Dataset will be used for training and the remaining 20% of the data will be used for testing the prediction of the trained model.

```
# Load the data and replace the '..' with nan
ts_df = pd.read_excel('MRR_zinc_metal_nickel_wire.xlsx')

ts_df = ts_df.astype(float)

display(ts_df)
```

	Ip	Ton	Toff	Wp	Wf	Wt	Sv	Sf	MRR
0	1.0	20.0	40.0	11.0	2.0	7.0	15.0	95.0	1.400
1	1.0	20.0	45.0	12.0	3.0	8.0	20.0	100.0	1.200
2	1.0	20.0	50.0	13.0	4.0	9.0	25.0	105.0	0.760
3	1.0	23.0	40.0	13.0	2.0	8.0	20.0	105.0	1.534
4	1.0	23.0	45.0	11.0	3.0	9.0	25.0	95.0	1.263
5	1.0	23.0	50.0	12.0	4.0	7.0	15.0	100.0	1.040
6	1.0	26.0	40.0	13.0	3.0	7.0	25.0	100.0	1.560
7	1.0	26.0	45.0	11.0	4.0	8.0	15.0	105.0	1.590
8	1.0	26.0	50.0	12.0	2.0	9.0	20.0	95.0	1.070
9	2.0	20.0	40.0	11.0	4.0	9.0	20.0	100.0	1.380
10	2.0	20.0	45.0	12.0	2.0	7.0	25.0	105.0	1.050
11	2.0	20.0	50.0	13.0	3.0	8.0	15.0	95.0	1.030
12	2.0	23.0	40.0	12.0	3.0	9.0	15.0	105.0	1.740
13	2.0	23.0	45.0	13.0	4.0	7.0	20.0	95.0	1.400
14	2.0	23.0	50.0	11.0	2.0	8.0	25.0	100.0	0.920
15	2.0	26.0	40.0	12.0	4.0	8.0	25.0	95.0	1.800
16	2.0	26.0	45.0	13.0	2.0	9.0	15.0	100.0	1.680
17	2.0	26.0	50.0	11.0	3.0	7.0	20.0	105.0	1.360

Fig. 4.18 Instructions to import MRR_Z_N

Figure 4.42 shows that 80% of the data from the SR Dataset will be used for training and the remaining 20% of the data will be used for testing the prediction of the trained model.

Figure 4.43 shows that 80% of the data from the MRR_B Dataset will be used for training and the remaining 20% of the data will be used for testing the prediction of the trained model.

Figure 4.44 shows that 80% of the data from the MRR_B_N Dataset will be used for training and the remaining 20% of the data will be used for testing the prediction of the trained model.

```
# Load the data and replace the '..' with nan
ts_df = pd.read_excel('SR_zinc_metal_nickel_wire.xlsx')

ts_df = ts_df.astype(float)

display(ts_df)
```

	Ip	Ton	Toff	Wp	Wf	Wt	Sv	Sf	SR
0	1.0	20.0	40.0	11.0	2.0	7.0	15.0	95.0	2.1445
1	1.0	20.0	45.0	12.0	3.0	8.0	20.0	100.0	2.4829
2	1.0	20.0	50.0	13.0	4.0	9.0	25.0	105.0	2.4520
3	1.0	23.0	40.0	13.0	2.0	8.0	20.0	105.0	2.5045
4	1.0	23.0	45.0	11.0	3.0	9.0	25.0	95.0	2.3218
5	1.0	23.0	50.0	12.0	4.0	7.0	15.0	100.0	2.5483
6	1.0	26.0	40.0	13.0	3.0	7.0	25.0	100.0	2.5124
7	1.0	26.0	45.0	11.0	4.0	8.0	15.0	105.0	2.2798
8	1.0	26.0	50.0	12.0	2.0	9.0	20.0	95.0	2.4732
9	2.0	20.0	40.0	11.0	4.0	9.0	20.0	100.0	2.2115
10	2.0	20.0	45.0	12.0	2.0	7.0	25.0	105.0	2.0818
11	2.0	20.0	50.0	13.0	3.0	8.0	15.0	95.0	2.2525
12	2.0	23.0	40.0	12.0	3.0	9.0	15.0	105.0	2.4889
13	2.0	23.0	45.0	13.0	4.0	7.0	20.0	95.0	2.4111
14	2.0	23.0	50.0	11.0	2.0	8.0	25.0	100.0	2.3660
15	2.0	26.0	40.0	12.0	4.0	8.0	25.0	95.0	2.4423
16	2.0	26.0	45.0	13.0	2.0	9.0	15.0	100.0	2.8251
17	2.0	26.0	50.0	11.0	3.0	7.0	20.0	105.0	2.2223

Fig. 4.19 Instructions to import SR_Z_N

Fig. 4.20 Instructions to import libraries for data visualization and analysis

```
import seaborn as sns
import pandas as pd
import matplotlib.pyplot as plt
```

	Ip(A)	Ton	TOW	MRR
Ip(A)	1.000000	0.000000	0.000000	0.978496
Ton	0.000000	1.000000	0.000000	0.035013
TOW	0.000000	0.000000	1.000000	0.124708
MRR	0.978496	0.035013	0.124708	1.000000

```
<matplotlib.axes._subplots.AxesSubplot at 0x2146eea23c8>
```

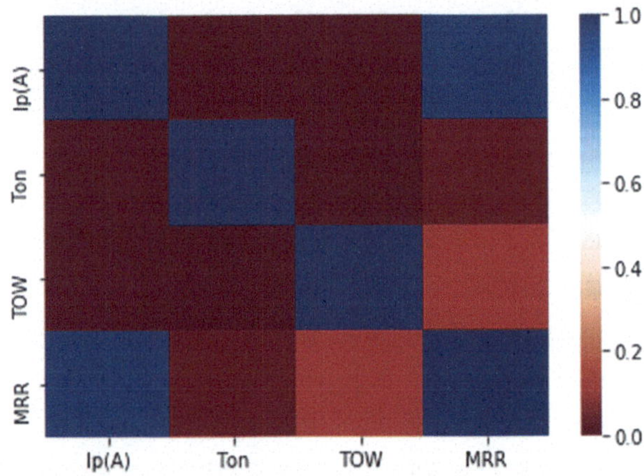

Fig. 4.21 MRR co-relation matrix

Figure 4.45 shows that 80% of the data from the SR_B_N Dataset will be used for training and the remaining 20% of the data will be used for testing the prediction of the trained model.

Figure 4.46 shows that 80% of the data from the MRR_Z Dataset will be used for training and the remaining 20% of the data will be used for testing the prediction of the trained model.

Figure 4.47 shows that 80% of the data from the MRR_Z_N Dataset will be used for training and the remaining 20% of the data will be used for testing the prediction of the trained model.

	Ip(A)	Ton	TOW	OC
Ip(A)	1.000000	0.000000	0.000000	0.844938
Ton	0.000000	1.000000	0.000000	0.372076
TOW	0.000000	0.000000	1.000000	0.020019
OC	0.844938	0.372076	0.020019	1.000000

```
<matplotlib.axes._subplots.AxesSubplot at 0x2b729d8e3c8>
```

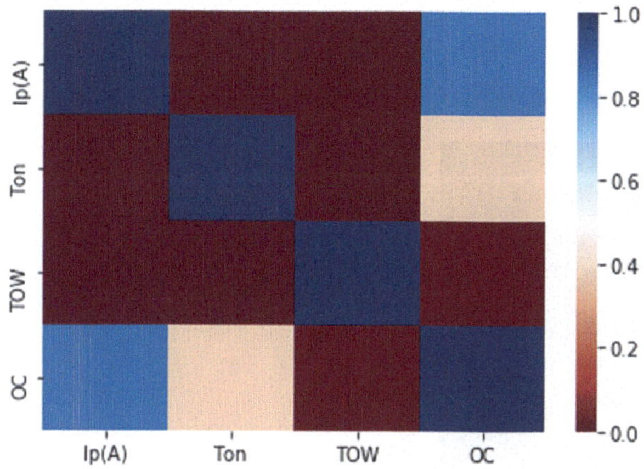

Fig. 4.22 OC co-relation matrix

Figure 4.48 shows that 80% of the data from the SR_Z_N Dataset will be used for training and the remaining 20% of the data will be used for testing the prediction of the trained model.

Once the model is trained, we use the trained model to predict the data in the following manner with the help of the following instructions (Fig 4.49).

The predictions for each dataset are observed as shown in the following Tables 4.9, 4.10, 4.11, 4.12, 4.13, 4.14, 4.15, 4.16, 4.17 are mentioned below.

From Fig. 4.50 it is observed that the actual value is quite in agreement with the predicted values of MRR.

```
# calculate the correlation matrix
corr = ts_df.corr()
# display the correlation matrix
display(corr)
# plot the correlation heatmap
sns.heatmap(corr, xticklabels=corr.columns, yticklabels=corr.columns, cmap='RdBu')
```

	Ip(A)	Ton	TOW	SR
Ip(A)	1.000000	0.000000	0.000000	0.882217
Ton	0.000000	1.000000	0.000000	0.255488
TOW	0.000000	0.000000	1.000000	0.045167
SR	0.882217	0.255488	0.045167	1.000000

```
<matplotlib.axes._subplots.AxesSubplot at 0x165e7325470>
```

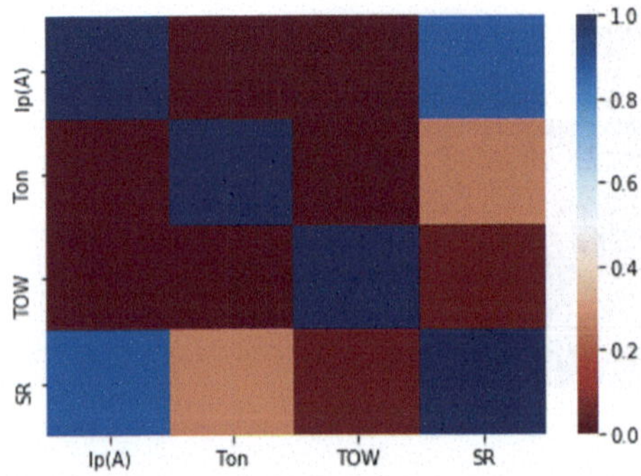

Fig. 4.23 SR co-relation matrix

From Fig. 4.51 it is observed that the actual value is quite in agreement with the predicted values of OC.

From Fig. 4.52 it is observed that the actual value is quite in agreement with the predicted values of SR.

From Fig. 4.53 it is observed that the actual value is quite in agreement with the predicted values of MRR_B.

From Fig. 4.54 it is observed that the actual value is quite in agreement with the predicted values of MRR_B_N.

	Ip (A)	Ton (μs)	Toff (μs)	WF (m/min)	WT (N)	SV (V)	SF (mm/min)	MRR (mm2/min) Brass wire
Ip (A)	1.00000	0.000000	0.000000	0.000000	0.000000	0.000000	0.000000	0.412740
Ton (μs)	0.00000	1.000000	0.000000	0.000000	0.000000	0.000000	0.000000	0.552892
Toff (μs)	0.00000	0.000000	1.000000	0.000000	0.000000	0.000000	0.000000	-0.173766
WF (m/min)	0.00000	0.000000	0.000000	1.000000	0.000000	0.000000	0.000000	0.071086
WT (N)	0.00000	0.000000	0.000000	0.000000	1.000000	0.000000	0.000000	-0.047391
SV (V)	0.00000	0.000000	0.000000	0.000000	0.000000	1.000000	0.000000	-0.315938
SF (mm/min)	0.00000	0.000000	0.000000	0.000000	0.000000	0.000000	1.000000	0.086883
MRR (mm2/min) Brass wire	0.41274	0.552892	-0.173766	0.071086	-0.047391	-0.315938	0.086883	1.000000

```
<matplotlib.axes._subplots.AxesSubplot at 0x1dfd1f474e0>
```

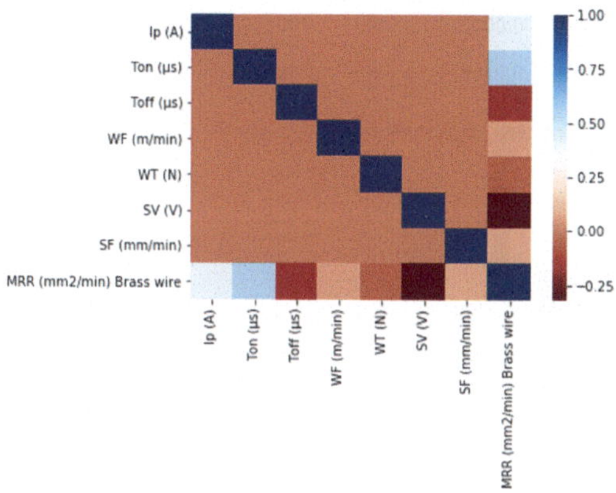

Fig. 4.24 MRR_B co-relation matrix

From Fig. 4.55 it is observed that the actual value is quite in agreement with the predicted values of SR_B_N.

From Fig. 4.56 it is observed that the actual value is quite in agreement with the predicted values of MRR_Z.

From Fig. 4.57 it is observed that the actual value is quite in agreement with the predicted values of MRR_Z_N.

From Fig. 4.58 it is observed that the actual value is quite in agreement with the predicted values of SR_Z_N.

	Ip	Ton	Toff	WP	Wf	Wt	SV	SF	MRR
Ip	1.000000	0.000000	0.000000	0.000000	0.000000	0.00000	0.000000	0.00000	0.275461
Ton	0.000000	1.000000	0.000000	0.000000	0.000000	0.00000	0.000000	0.00000	0.300127
Toff	0.000000	0.000000	1.000000	0.000000	0.000000	0.00000	0.000000	0.00000	-0.801799
WP	0.000000	0.000000	0.000000	1.000000	0.000000	0.00000	0.000000	0.00000	0.173066
Wf	0.000000	0.000000	0.000000	0.000000	1.000000	0.00000	0.000000	0.00000	-0.056958
Wt	0.000000	0.000000	0.000000	0.000000	0.000000	1.00000	0.000000	0.00000	0.124870
SV	0.000000	0.000000	0.000000	0.000000	0.000000	0.00000	1.000000	0.00000	-0.197164
SF	0.000000	0.000000	0.000000	0.000000	0.000000	0.00000	0.000000	1.00000	-0.030670
MRR	0.275461	0.300127	-0.801799	0.173066	-0.056958	0.12487	-0.197164	-0.03067	1.000000

```
<matplotlib.axes._subplots.AxesSubplot at 0x1cb98604668>
```

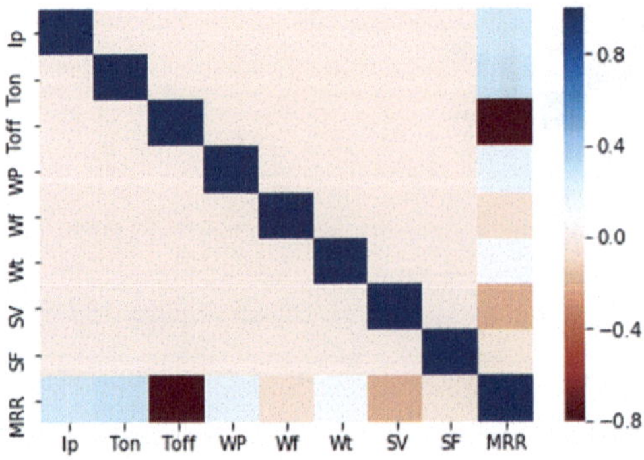

Fig. 4.25 MRR_B_N co-relation matrix

	Ip	Ton	Toff	WP	Wf	Wt	SV	SF	SR
Ip	1.000000	0.000000	0.000000	0.000000	0.000000	0.000000	0.00000	0.00000	0.366499
Ton	0.000000	1.000000	0.000000	0.000000	0.000000	0.000000	0.00000	0.00000	0.380956
Toff	0.000000	0.000000	1.000000	0.000000	0.000000	0.000000	0.00000	0.00000	-0.114395
WP	0.000000	0.000000	0.000000	1.000000	0.000000	0.000000	0.00000	0.00000	0.175613
Wf	0.000000	0.000000	0.000000	0.000000	1.000000	0.000000	0.00000	0.00000	-0.023747
Wt	0.000000	0.000000	0.000000	0.000000	0.000000	1.000000	0.00000	0.00000	-0.369364
SV	0.000000	0.000000	0.000000	0.000000	0.000000	0.000000	1.00000	0.00000	-0.292140
SF	0.000000	0.000000	0.000000	0.000000	0.000000	0.000000	0.00000	1.00000	-0.215590
SR	0.366499	0.380956	-0.114395	0.175613	-0.023747	-0.369364	-0.29214	-0.21559	1.000000

```
<matplotlib.axes._subplots.AxesSubplot at 0x1abf05816a0>
```

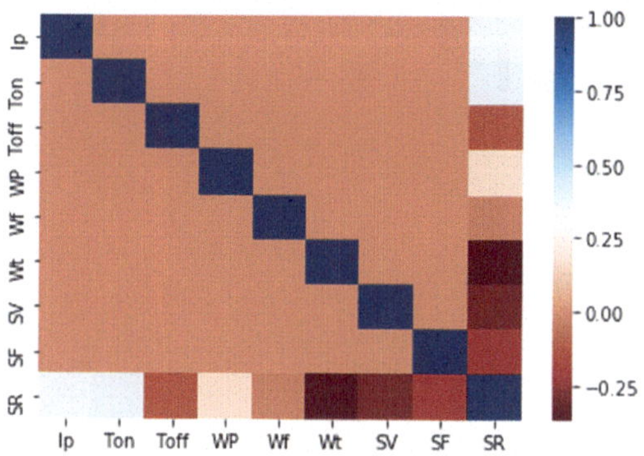

Fig. 4.26 SR_B_N co-relation matrix

	Ip (A)	Ton (µs)	Toff (µs)	WF (m/min)	WT (N)	SV (V)	SF (mm/min)	MRR (mm2/min)Zin coated wire
Ip (A)	1.000000	0.000000	0.000000	0.000000	0.000000	0.000000	0.000000	-0.056008
Ton (µs)	0.000000	1.000000	0.000000	0.000000	0.000000	0.000000	0.000000	0.698428
Toff (µs)	0.000000	0.000000	1.000000	0.000000	0.000000	0.000000	0.000000	-0.311798
WF (m/min)	0.000000	0.000000	0.000000	1.000000	0.000000	0.000000	0.000000	-0.087304
WT (N)	0.000000	0.000000	0.000000	0.000000	1.000000	0.000000	0.000000	0.193315
SV (V)	0.000000	0.000000	0.000000	0.000000	0.000000	1.000000	0.000000	-0.405338
SF (mm/min)	0.000000	0.000000	0.000000	0.000000	0.000000	0.000000	1.000000	0.212023
MRR (mm2/min)Zin coated wire	-0.056008	0.698428	-0.311798	-0.087304	0.193315	-0.405338	0.212023	1.000000

```
<matplotlib.axes._subplots.AxesSubplot at 0x19fb3963518>
```

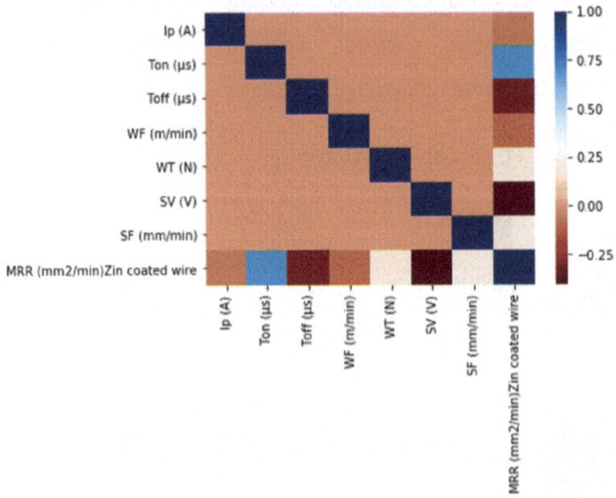

Fig. 4.27 MRR_Z co-relation matrix

	Ip	Ton	Toff	Wp	Wf	Wt	Sv	Sf	MRR
Ip	1.000000	0.000000	0.000000	0.000000	0.000000	0.000000	0.000000	0.000000	0.180293
Ton	0.000000	1.000000	0.000000	0.000000	0.000000	0.000000	0.000000	0.000000	0.524518
Toff	0.000000	0.000000	1.000000	0.000000	0.000000	0.000000	0.000000	0.000000	-0.757272
Wp	0.000000	0.000000	0.000000	1.000000	0.000000	0.000000	0.000000	0.000000	0.011942
Wf	0.000000	0.000000	0.000000	0.000000	1.000000	0.000000	0.000000	0.000000	0.073994
Wt	0.000000	0.000000	0.000000	0.000000	0.000000	1.000000	0.000000	0.000000	0.019435
Sv	0.000000	0.000000	0.000000	0.000000	0.000000	0.000000	1.000000	0.000000	-0.263898
Sf	0.000000	0.000000	0.000000	0.000000	0.000000	0.000000	0.000000	1.000000	0.016625
MRR	0.180293	0.524518	-0.757272	0.011942	0.073994	0.019435	-0.263898	0.016625	1.000000

```
<matplotlib.axes._subplots.AxesSubplot at 0x1f9b01f35c0>
```

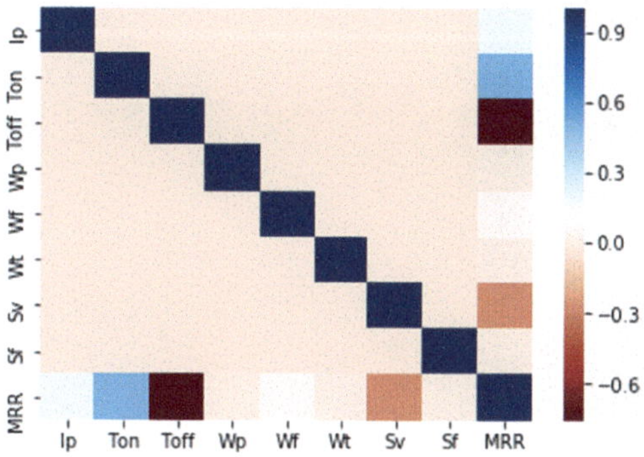

Fig. 4.28 MRR_Z_N co-relation matrix

	Ip	Ton	Toff	Wp	Wf	Wt	Sv	Sf	SR
Ip	1.000000	0.000000	0.000000	0.000000	0.000000	0.000000	0.000000	0.000000	-0.135769
Ton	0.000000	1.000000	0.000000	0.000000	0.000000	0.000000	0.000000	0.000000	0.449589
Toff	0.000000	0.000000	1.000000	0.000000	0.000000	0.000000	0.000000	0.000000	0.004059
Wp	0.000000	0.000000	0.000000	1.000000	0.000000	0.000000	0.000000	0.000000	0.561717
Wf	0.000000	0.000000	0.000000	0.000000	1.000000	0.000000	0.000000	0.000000	-0.019935
Wt	0.000000	0.000000	0.000000	0.000000	0.000000	1.000000	0.000000	0.000000	0.339052
Sv	0.000000	0.000000	0.000000	0.000000	0.000000	0.000000	1.000000	0.000000	-0.144359
Sf	0.000000	0.000000	0.000000	0.000000	0.000000	0.000000	0.000000	1.000000	-0.006406
SR	-0.135769	0.449589	0.004059	0.561717	-0.019935	0.339052	-0.144359	-0.006406	1.000000

```
<matplotlib.axes._subplots.AxesSubplot at 0x2a30d1a6208>
```

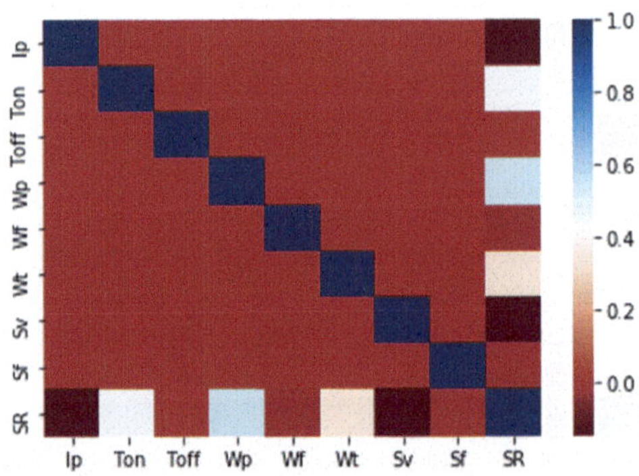

Fig. 4.29 SR_Z_N co-relation matrix

```
# get the summary
desc_df = ts_df.describe()
# add the standard deviation metric
desc_df.loc['+3_std'] = desc_df.loc['mean'] + (desc_df.loc['std'] * 3)
desc_df.loc['-3_std'] = desc_df.loc['mean'] - (desc_df.loc['std'] * 3)
# display it
desc_df
```

	Ip(A)	Ton	TOW	MRR
count	27.000000	27.000000	27.000000	27.000000
mean	5.000000	100.000000	80.000000	9.256222
std	2.496151	41.602515	8.320503	6.502541
min	2.000000	50.000000	70.000000	1.312500
25%	2.000000	50.000000	70.000000	2.137500
50%	5.000000	100.000000	80.000000	8.796800
75%	8.000000	150.000000	90.000000	15.214600
max	8.000000	150.000000	90.000000	20.467000
+3_std	12.488453	224.807544	104.961509	28.763844
-3_std	-2.488453	-24.807544	55.038491	-10.251400

```
# filter the data frame to remove the values exceeding 3 standard deviations
econ_remove_df = ts_df[(np.abs(stats.zscore(ts_df)) < 3).all(axis=1)]
# what rows were removed
ts_df.index.difference(econ_remove_df.index)
```

```
Int64Index([], dtype='int64')
```

Fig. 4.30 Instructions to preprocess the MRR Dataset

```
# get the summary
desc_df = ts_df.describe()
# add the standard deviation metric
desc_df.loc['+3_std'] = desc_df.loc['mean'] + (desc_df.loc['std'] * 3)
desc_df.loc['-3_std'] = desc_df.loc['mean'] - (desc_df.loc['std'] * 3)
# display it
desc_df
```

	Ip(A)	Ton	TOW	OC
count	27.000000	27.000000	27.000000	27.000000
mean	5.000000	100.000000	80.000000	0.124407
std	2.496151	41.602515	8.320503	0.066963
min	2.000000	50.000000	70.000000	0.012000
25%	2.000000	50.000000	70.000000	0.083000
50%	5.000000	100.000000	80.000000	0.106000
75%	8.000000	150.000000	90.000000	0.185500
max	8.000000	150.000000	90.000000	0.226000
+3_std	12.488453	224.807544	104.961509	0.325296
-3_std	-2.488453	-24.807544	55.038491	-0.076481

Fig. 4.31 Instructions to preprocess the OC Dataset

	Ip(A)	Ton	TOW	SR
count	27.000000	27.000000	27.000000	27.000000
mean	5.000000	100.000000	80.000000	6.840926
std	2.496151	41.602515	8.320503	2.029471
min	2.000000	50.000000	70.000000	4.067000
25%	2.000000	50.000000	70.000000	4.968500
50%	5.000000	100.000000	80.000000	6.867000
75%	8.000000	150.000000	90.000000	7.898500
max	8.000000	150.000000	90.000000	10.130000
+3_std	12.488453	224.807544	104.961509	12.929338
-3_std	-2.488453	-24.807544	55.038491	0.752514

```
# filter the data frame to remove the values exceeding 3 standard deviations
econ_remove_df = ts_df[(np.abs(stats.zscore(ts_df)) < 3).all(axis=1)]
# what rows were removed
ts_df.index.difference(econ_remove_df.index)
```

```
Int64Index([], dtype='int64')
```

Fig. 4.32 Instructions to preprocess the SR Dataset

```
# get the summary
desc_df = ts_df.describe()
# add the standard deviation metric
desc_df.loc['+3_std'] = desc_df.loc['mean'] + (desc_df.loc['std'] * 3)
desc_df.loc['-3_std'] = desc_df.loc['mean'] - (desc_df.loc['std'] * 3)
# display it
desc_df
```

	Ip (A)	Ton (μs)	Toff (μs)	WF (m/min)	WT (N)	SV (V)	SF (mm/min)	MRR (mm2/min) Brass wire
count	18.000000	18.000000	18.000000	18.000000	18.000000	18.000000	18.000000	18.000000
mean	1.500000	25.000000	45.000000	6.000000	8.000000	20.000000	2100.000000	0.661111
std	0.514496	2.520504	4.200840	0.840168	1.680336	4.200840	8.401681	0.088643
min	1.000000	22.000000	40.000000	5.000000	6.000000	15.000000	2090.000000	0.530000
25%	1.000000	22.000000	40.000000	5.000000	6.000000	15.000000	2090.000000	0.590000
50%	1.500000	25.000000	45.000000	6.000000	8.000000	20.000000	2100.000000	0.635000
75%	2.000000	28.000000	50.000000	7.000000	10.000000	25.000000	2110.000000	0.712500
max	2.000000	28.000000	50.000000	7.000000	10.000000	25.000000	2110.000000	0.830000
+3_std	3.043487	32.561512	57.602521	8.520504	13.041008	32.602521	2125.205042	0.927039
-3_std	-0.043487	17.438488	32.397479	3.479496	2.958992	7.397479	2074.794958	0.395183

```
# filter the data frame to remove the values exceeding 3 standard deviations
econ_remove_df = ts_df[(np.abs(stats.zscore(ts_df)) < 3).all(axis=1)]
# what rows were removed
ts_df.index.difference(econ_remove_df.index)
```

```
Int64Index([], dtype='int64')
```

Fig. 4.33 Instructions to preprocess the MRR_B Dataset

```
# get the summary
desc_df = ts_df.describe()
# add the standard deviation metric
desc_df.loc['+3_std'] = desc_df.loc['mean'] + (desc_df.loc['std'] * 3)
desc_df.loc['-3_std'] = desc_df.loc['mean'] - (desc_df.loc['std'] * 3)
# display it
desc_df
```

	Ip	Ton	Toff	WP	Wf	Wt	SV	SF	MRR
count	18.000000	18.000000	18.000000	18.000000	18.000000	18.000000	18.000000	18.000000	18.000000
mean	1.500000	23.000000	45.000000	12.000000	3.000000	8.000000	20.000000	100.000000	1.216667
std	0.514496	2.520504	4.200840	0.840168	0.840168	0.840168	4.200840	4.200840	0.319595
min	1.000000	20.000000	40.000000	11.000000	2.000000	7.000000	15.000000	95.000000	0.720000
25%	1.000000	20.000000	40.000000	11.000000	2.000000	7.000000	15.000000	95.000000	0.930000
50%	1.500000	23.000000	45.000000	12.000000	3.000000	8.000000	20.000000	100.000000	1.245000
75%	2.000000	26.000000	50.000000	13.000000	4.000000	9.000000	25.000000	105.000000	1.352500
max	2.000000	26.000000	50.000000	13.000000	4.000000	9.000000	25.000000	105.000000	1.820000
+3_std	3.043487	30.561512	57.602521	14.520504	5.520504	10.520504	32.602521	112.602521	2.175453
-3_std	-0.043487	15.438488	32.397479	9.479496	0.479496	5.479496	7.397479	87.397479	0.257881

Fig. 4.34 Instructions to preprocess the MRR_B_N Dataset

```
# get the summary
desc_df = ts_df.describe()
# add the standard deviation metric
desc_df.loc['+3_std'] = desc_df.loc['mean'] + (desc_df.loc['std'] * 3)
desc_df.loc['-3_std'] = desc_df.loc['mean'] - (desc_df.loc['std'] * 3)
# display it
desc_df
```

	Ip	Ton	Toff	WP	Wf	Wt	SV	SF	SR
count	18.000000	18.000000	18.000000	18.000000	18.000000	18.000000	18.000000	18.000000	18.000000
mean	1.500000	23.000000	45.000000	12.000000	3.000000	8.000000	20.000000	100.000000	2.329794
std	0.514496	2.520504	4.200840	0.840168	0.840168	0.840168	4.200840	4.200840	0.187222
min	1.000000	20.000000	40.000000	11.000000	2.000000	7.000000	15.000000	95.000000	1.882400
25%	1.000000	20.000000	40.000000	11.000000	2.000000	7.000000	15.000000	95.000000	2.263900
50%	1.500000	23.000000	45.000000	12.000000	3.000000	8.000000	20.000000	100.000000	2.363100
75%	2.000000	26.000000	50.000000	13.000000	4.000000	9.000000	25.000000	105.000000	2.453575
max	2.000000	26.000000	50.000000	13.000000	4.000000	9.000000	25.000000	105.000000	2.631600
+3_std	3.043487	30.561512	57.602521	14.520504	5.520504	10.520504	32.602521	112.602521	2.891459
-3_std	-0.043487	15.438488	32.397479	9.479496	0.479496	5.479496	7.397479	87.397479	1.768130

```
# filter the data frame to remove the values exceeding 3 standard deviations
econ_remove_df = ts_df[(np.abs(stats.zscore(ts_df)) < 3).all(axis=1)]
# what rows were removed
ts_df.index.difference(econ_remove_df.index)
```

```
Int64Index([], dtype='int64')
```

Fig. 4.35 Instructions to preprocess the SR_B_N Dataset

```
# get the summary
desc_df = ts_df.describe()
# add the standard deviation metric
desc_df.loc['+3_std'] = desc_df.loc['mean'] + (desc_df.loc['std'] * 3)
desc_df.loc['-3_std'] = desc_df.loc['mean'] - (desc_df.loc['std'] * 3)
# display it
desc_df
```

	Ip (A)	Ton (μs)	Toff (μs)	WF (m/min)	WT (N)	SV (V)	SF (mm/min)	MRR (mm2/min)Zin coated wire
count	18.000000	18.000000	18.000000	18.000000	18.000000	18.000000	18.000000	18.000000
mean	1.500000	25.000000	45.000000	6.000000	8.000000	20.000000	2100.000000	0.839444
std	0.514496	2.520504	4.200840	0.840168	1.680336	4.200840	8.401681	0.112274
min	1.000000	22.000000	40.000000	5.000000	6.000000	15.000000	2090.000000	0.670000
25%	1.000000	22.000000	40.000000	5.000000	6.000000	15.000000	2090.000000	0.752500
50%	1.500000	25.000000	45.000000	6.000000	8.000000	20.000000	2100.000000	0.860000
75%	2.000000	28.000000	50.000000	7.000000	10.000000	25.000000	2110.000000	0.880000
max	2.000000	28.000000	50.000000	7.000000	10.000000	25.000000	2110.000000	1.020000
+3_std	3.043487	32.561512	57.602521	8.520504	13.041008	32.602521	2125.205042	1.176268
-3_std	-0.043487	17.438488	32.397479	3.479496	2.958992	7.397479	2074.794958	0.502621

```
# filter the data frame to remove the values exceeding 3 standard deviations
econ_remove_df = ts_df[(np.abs(stats.zscore(ts_df)) < 3).all(axis=1)]
# what rows were removed
ts_df.index.difference(econ_remove_df.index)
```

```
Int64Index([], dtype='int64')
```

Fig. 4.36 Instructions to preprocess the MRR_Z Dataset

```
# get the summary
desc_df = ts_df.describe()
# add the standard deviation metric
desc_df.loc['+3_std'] = desc_df.loc['mean'] + (desc_df.loc['std'] * 3)
desc_df.loc['-3_std'] = desc_df.loc['mean'] - (desc_df.loc['std'] * 3)
# display it
desc_df
```

	Ip	Ton	Toff	Wp	Wf	Wt	Sv	Sf	MRR
count	18.000000	18.000000	18.000000	18.000000	18.000000	18.000000	18.000000	18.000000	18.000000
mean	1.500000	23.000000	45.000000	12.000000	3.000000	8.000000	20.000000	100.000000	1.320944
std	0.514496	2.520504	4.200840	0.840168	0.840168	0.840168	4.200840	4.200840	0.299001
min	1.000000	20.000000	40.000000	11.000000	2.000000	7.000000	15.000000	95.000000	0.760000
25%	1.000000	20.000000	40.000000	11.000000	2.000000	7.000000	15.000000	95.000000	1.055000
50%	1.500000	23.000000	45.000000	12.000000	3.000000	8.000000	20.000000	100.000000	1.370000
75%	2.000000	26.000000	50.000000	13.000000	4.000000	9.000000	25.000000	105.000000	1.553500
max	2.000000	26.000000	50.000000	13.000000	4.000000	9.000000	25.000000	105.000000	1.800000
+3_std	3.043487	30.561512	57.602521	14.520504	5.520504	10.520504	32.602521	112.602521	2.217948
-3_std	-0.043487	15.438488	32.397479	9.479496	0.479496	5.479496	7.397479	87.397479	0.423941

```
# filter the data frame to remove the values exceeding 3 standard deviations
econ_remove_df = ts_df[(np.abs(stats.zscore(ts_df)) < 3).all(axis=1)]
# what rows were removed
ts_df.index.difference(econ_remove_df.index)
```

```
Int64Index([], dtype='int64')
```

Fig. 4.37 Instructions to preprocess the MRR_Z_N Dataset

```
# get the summary
desc_df = ts_df.describe()
# add the standard deviation metric
desc_df.loc['+3_std'] = desc_df.loc['mean'] + (desc_df.loc['std'] * 3)
desc_df.loc['-3_std'] = desc_df.loc['mean'] - (desc_df.loc['std'] * 3)
# display it
desc_df
```

	Ip	Ton	Toff	Wp	Wf	Wt	Sv	Sf	SR
count	18.000000	18.000000	18.000000	18.000000	18.000000	18.000000	18.000000	18.000000	18.000000
mean	1.500000	23.000000	45.000000	12.000000	3.000000	8.000000	20.000000	100.000000	2.390050
std	0.514496	2.520504	4.200840	0.840168	0.840168	0.840168	4.200840	4.200840	0.175958
min	1.000000	20.000000	40.000000	11.000000	2.000000	7.000000	15.000000	95.000000	2.081800
25%	1.000000	20.000000	40.000000	11.000000	2.000000	7.000000	15.000000	95.000000	2.259325
50%	1.500000	23.000000	45.000000	12.000000	3.000000	8.000000	20.000000	100.000000	2.426700
75%	2.000000	26.000000	50.000000	13.000000	4.000000	9.000000	25.000000	105.000000	2.487400
max	2.000000	26.000000	50.000000	13.000000	4.000000	9.000000	25.000000	105.000000	2.825100
+3_std	3.043487	30.561512	57.602521	14.520504	5.520504	10.520504	32.602521	112.602521	2.917925
-3_std	-0.043487	15.438488	32.397479	9.479496	0.479496	5.479496	7.397479	87.397479	1.862175

```
# filter the data frame to remove the values exceeding 3 standard deviations
econ_remove_df = ts_df[(np.abs(stats.zscore(ts_df)) < 3).all(axis=1)]
# what rows were removed
ts_df.index.difference(econ_remove_df.index)
```

```
Int64Index([], dtype='int64')
```

Fig. 4.38 Instructions to preprocess the SR_Z_N Dataset

```
# define our input variable (X) & output variable (Y)
X = ts_df.iloc[:, :-1]
Y = ts_df.iloc[:, 3]
Y

# Split X and y into X_
X_train, X_test, y_train, y_test = train_test_split(X, Y, test_size=.20, random_state=0)
#X_train, X_test, y_train, y_test = model_selection.train_test_split(X, y, train_size=0.75,test_size=0.25, random_state=101)
print ("X_train: ", X_train)
print ("y_train: ", y_train)
print("X_test: ", X_test)
print ("y_test: ", y_test)

# create a Linear Regression model object
regression_model = LinearRegression()

# pass through the X_train & y_train data set
regression_model.fit(X_train, y_train)
```

Fig. 4.39 Instructions to split the dataset into training and test data

```
X_train:      Ip(A)    Ton    TOW
23    8.0  100.0  90.0
13    5.0  100.0  80.0
19    8.0   50.0  80.0
20    8.0   50.0  90.0
16    5.0  150.0  80.0
1     2.0   50.0  80.0
10    5.0   50.0  80.0
26    8.0  150.0  90.0
25    8.0  150.0  80.0
8     2.0  150.0  90.0
6     2.0  150.0  70.0
4     2.0  100.0  80.0
18    8.0   50.0  70.0
9     5.0   50.0  70.0
7     2.0  150.0  80.0
22    8.0  100.0  80.0
3     2.0  100.0  70.0
0     2.0   50.0  70.0
21    8.0  100.0  70.0
15    5.0  150.0  70.0
12    5.0  100.0  70.0
y_train:  23     20.4670
13     9.2250
19    17.1000
20    17.5167
16     9.1750
1      2.3063
10     8.3167
26    19.3500
25    16.9875
8      2.1250
6      1.9000
4      2.0625
18    14.2417
9      7.3500
7      2.1188
22    18.8670
3      1.7938
0      1.5438
21    15.8417
15     7.6580
12     8.1917
Name: MRR, dtype: float64
X_test:       Ip(A)    Ton    TOW
2     2.0   50.0  90.0
24    8.0  150.0  70.0
14    5.0  100.0  90.0
17    5.0  150.0  90.0
5     2.0  100.0  90.0
11    5.0   50.0  90.0

y_test:  2      1.3125
24    14.5875
14     9.4250
17     9.5080
5      2.1500
11     8.7968
Name: MRR, dtype: float64

LinearRegression(copy_X=True, fit_intercept=True, n_jobs=None,
         normalize=False)
```

Fig. 4.40 Instructions to split MRR Dataset into 20% test and 80% training data

```
X_train:       Ip(A)     Ton    TOW
23    8.0  100.0  90.0
13    5.0  100.0  80.0
19    8.0   50.0  80.0
20    8.0   50.0  90.0
16    5.0  150.0  80.0
1     2.0   50.0  80.0
10    5.0   50.0  80.0
26    8.0  150.0  90.0
25    8.0  150.0  80.0
8     2.0  150.0  90.0
6     2.0  150.0  70.0
4     2.0  100.0  80.0
18    8.0   50.0  70.0
9     5.0   50.0  70.0
7     2.0  150.0  80.0
22    8.0  100.0  80.0
3     2.0  100.0  70.0
0     2.0   50.0  70.0
21    8.0  100.0  70.0
15    5.0  150.0  70.0
12    5.0  100.0  70.0
y_train:  23    0.200
13    0.144
19    0.174
20    0.158
16    0.195
1     0.018
10    0.130
26    0.226
25    0.204
8     0.096
6     0.096
4     0.056
18    0.181
9     0.106
7     0.086
22    0.225
3     0.086
0     0.012
21    0.189
15    0.096
12    0.082
Name: OC, dtype: float64
X_test:       Ip(A)     Ton    TOW
2     2.0   50.0  90.0
24    8.0  150.0  70.0
14    5.0  100.0  90.0
17    5.0  150.0  90.0
5     2.0  100.0  90.0
11    5.0   50.0  90.0
y_test:  2     0.016
24    0.201
14    0.084
17    0.182
5     0.068
11    0.048
Name: OC, dtype: float64

LinearRegression(copy_X=True, fit_intercept=True, n_jobs=None,
         normalize=False)
```

Fig. 4.41 Instructions to split OC Dataset into 20% test and 80% training data

```
X_train:      Ip(A)    Ton   TOW
23    8.0  100.0  90.0
13    5.0  100.0  80.0
19    8.0   50.0  80.0
20    8.0   50.0  90.0
16    5.0  150.0  80.0
1     2.0   50.0  80.0
10    5.0   50.0  80.0
26    8.0  150.0  90.0
25    8.0  150.0  80.0
8     2.0  150.0  90.0
6     2.0  150.0  70.0
4     2.0  100.0  80.0
18    8.0   50.0  70.0
9     5.0   50.0  70.0
7     2.0  150.0  80.0
22    8.0  100.0  80.0
3     2.0  100.0  70.0
0     2.0   50.0  70.0
21    8.0  100.0  70.0
15    5.0  150.0  70.0
12    5.0  100.0  70.0
y_train:  23     10.130
13      7.200
19      6.930
20      7.530
16      7.800
1       5.330
10      6.530
26     10.000
25      9.800
8       4.067
6       4.670
4       4.530
18      6.600
9       6.000
7       4.600
22     10.000
3       4.400
0       5.267
21      8.867
15      6.867
12      7.330
Name: SR, dtype: float64
X_test:      Ip(A)    Ton   TOW
2     2.0   50.0  90.0
24    8.0  150.0  70.0
14    5.0  100.0  90.0
17    5.0  150.0  90.0
5     2.0  100.0  90.0
11    5.0   50.0  90.0
y_test:  2      4.130
24     10.000
14      7.867
17      7.930
5       4.130
11      6.200
Name: SR, dtype: float64

LinearRegression(copy_X=True, fit_intercept=True, n_jobs=None,
         normalize=False)
```

Fig. 4.42 Instructions to split SR Dataset into 20% test and 80% training data

```
X_train:      Ip (A)  Ton (µs)  Toff (µs)  WF (m/min)  WT (N)  SV (V)  SF (mm/min)
14      2.0     25.0      50.0          5.0       8.0    25.0        2100.0
7       1.0     28.0      45.0          7.0       8.0    15.0        2110.0
15      2.0     28.0      40.0          7.0       8.0    25.0        2090.0
4       1.0     25.0      45.0          6.0      10.0    25.0        2090.0
1       1.0     22.0      45.0          6.0       8.0    20.0        2100.0
10      2.0     22.0      45.0          5.0       6.0    25.0        2110.0
0       1.0     22.0      40.0          5.0       6.0    15.0        2090.0
17      2.0     28.0      50.0          6.0       6.0    20.0        2110.0
16      2.0     28.0      45.0          5.0      10.0    15.0        2100.0
9       2.0     22.0      40.0          7.0      10.0    20.0        2100.0
8       1.0     28.0      50.0          5.0      10.0    20.0        2090.0
12      2.0     25.0      40.0          6.0      10.0    15.0        2110.0
11      2.0     22.0      50.0          6.0       8.0    15.0        2090.0
5       1.0     25.0      50.0          7.0       6.0    15.0        2100.0
y_train:  14    0.59
7      0.75
15     0.72
4      0.60
1      0.59
10     0.69
0      0.58
17     0.69
16     0.77
9      0.57
8      0.69
12     0.83
11     0.59
5      0.64
Name: MRR (mm2/min) Brass wire, dtype: float64
X_test:       Ip (A)  Ton (µs)  Toff (µs)  WF (m/min)  WT (N)  SV (V)  SF (mm/min)
6       1.0     28.0      40.0          6.0       6.0    25.0        2100.0
3       1.0     25.0      40.0          5.0       8.0    20.0        2110.0
13      2.0     25.0      45.0          7.0       6.0    20.0        2090.0
2       1.0     22.0      50.0          7.0      10.0    25.0        2110.0
y_test:  6     0.63
3      0.62
13     0.82
2      0.53
Name: MRR (mm2/min) Brass wire, dtype: float64

LinearRegression(copy_X=True, fit_intercept=True, n_jobs=None,
         normalize=False)
```

Fig. 4.43 Instructions to split MRR_B Dataset into 20% test and 80% training data

```
X_train:        Ip    Ton   Toff    WP    Wf    Wt     SV      SF
14  2.0  23.0  50.0  11.0  2.0  8.0  25.0  100.0
7   1.0  26.0  45.0  11.0  4.0  8.0  15.0  105.0
15  2.0  26.0  40.0  12.0  4.0  8.0  25.0   95.0
4   1.0  23.0  45.0  11.0  3.0  9.0  25.0   95.0
1   1.0  20.0  45.0  12.0  3.0  8.0  20.0  100.0
10  2.0  20.0  45.0  12.0  2.0  7.0  25.0  105.0
0   1.0  20.0  40.0  11.0  2.0  7.0  15.0   95.0
17  2.0  26.0  50.0  11.0  3.0  7.0  20.0  105.0
16  2.0  26.0  45.0  13.0  2.0  9.0  15.0  100.0
9   2.0  20.0  40.0  11.0  4.0  9.0  20.0  100.0
8   1.0  26.0  50.0  12.0  2.0  9.0  20.0   95.0
12  2.0  23.0  40.0  12.0  3.0  9.0  15.0  105.0
11  2.0  20.0  50.0  13.0  3.0  8.0  15.0   95.0
5   1.0  23.0  50.0  12.0  4.0  7.0  15.0  100.0
y_train:  14    0.89
7     1.29
15    1.75
4     1.12
1     1.09
10    1.20
0     1.32
17    0.85
16    1.82
9     1.33
8     0.89
12    1.65
11    0.93
5     0.93
Name: MRR, dtype: float64
X_test:        Ip    Ton   Toff    WP    Wf    Wt     SV      SF
6   1.0  26.0  40.0  13.0  3.0  7.0  25.0  100.0
3   1.0  23.0  40.0  13.0  2.0  8.0  20.0  105.0
13  2.0  23.0  45.0  13.0  4.0  7.0  20.0   95.0
2   1.0  20.0  50.0  13.0  4.0  9.0  25.0  105.0
y_test:  6     1.36
3     1.46
13    1.30
2     0.72
Name: MRR, dtype: float64

LinearRegression(copy_X=True, fit_intercept=True, n_jobs=None,
         normalize=False)
```

Fig. 4.44 Instructions to split MRR_B_N Dataset into 20% test and 80% training data

```
X_train:        Ip   Ton  Toff    WP   Wf   Wt    SV     SF
14  2.0  23.0  50.0  11.0  2.0  8.0  25.0  100.0
7   1.0  26.0  45.0  11.0  4.0  8.0  15.0  105.0
15  2.0  26.0  40.0  12.0  4.0  8.0  25.0   95.0
4   1.0  23.0  45.0  11.0  3.0  9.0  25.0   95.0
1   1.0  20.0  45.0  12.0  3.0  8.0  20.0  100.0
10  2.0  20.0  45.0  12.0  2.0  7.0  25.0  105.0
0   1.0  20.0  40.0  11.0  2.0  7.0  15.0   95.0
17  2.0  26.0  50.0  11.0  3.0  7.0  20.0  105.0
16  2.0  26.0  45.0  13.0  2.0  9.0  15.0  100.0
9   2.0  20.0  40.0  11.0  4.0  9.0  20.0  100.0
8   1.0  26.0  50.0  12.0  2.0  9.0  20.0   95.0
12  2.0  23.0  40.0  12.0  3.0  9.0  15.0  105.0
11  2.0  20.0  50.0  13.0  3.0  8.0  15.0   95.0
5   1.0  23.0  50.0  12.0  4.0  7.0  15.0  100.0
y_train:  14    2.3896
7     2.3882
15    2.6316
4     2.1456
1     2.4543
10    2.2858
0     2.2566
17    2.3337
16    2.4514
9     1.9849
8     2.2094
12    2.4876
11    2.4696
5     2.4008
Name: SR, dtype: float64
X_test:        Ip   Ton  Toff    WP   Wf   Wt    SV     SF
6   1.0  26.0  40.0  13.0  3.0  7.0  25.0  100.0
3   1.0  23.0  40.0  13.0  2.0  8.0  20.0  105.0
13  2.0  23.0  45.0  13.0  4.0  7.0  20.0   95.0
2   1.0  20.0  50.0  13.0  4.0  9.0  25.0  105.0
y_test:  6    2.3380
3     2.2927
13    2.5341
2     1.8824
Name: SR, dtype: float64

LinearRegression(copy_X=True, fit_intercept=True, n_jobs=None,
         normalize=False)
```

Fig. 4.45 Instructions to split SR_B_N Dataset into 20% test and 80% training data

```
X_train:      Ip (A)  Ton (μs)  Toff (μs)  WF (m/min)  WT (N)  SV (V)  SF (mm/min
14     2.0     25.0       50.0        5.0        8.0    25.0      2100.0
7      1.0     28.0       45.0        7.0        8.0    15.0      2110.0
15     2.0     28.0       40.0        7.0        8.0    25.0      2090.0
4      1.0     25.0       45.0        6.0       10.0    25.0      2090.0
1      1.0     22.0       45.0        6.0        8.0    20.0      2100.0
10     2.0     22.0       45.0        5.0        6.0    25.0      2110.0
0      1.0     22.0       40.0        5.0        6.0    15.0      2090.0
17     2.0     28.0       50.0        6.0        6.0    20.0      2110.0
16     2.0     28.0       45.0        5.0       10.0    15.0      2100.0
9      2.0     22.0       40.0        7.0       10.0    20.0      2100.0
8      1.0     28.0       50.0        5.0       10.0    20.0      2090.0
12     2.0     25.0       40.0        6.0       10.0    15.0      2110.0
11     2.0     22.0       50.0        6.0        8.0    15.0      2090.0
5      1.0     25.0       50.0        7.0        6.0    15.0      2100.0
y_train:  14    0.75
7      0.99
15     0.87
4      0.88
1      0.69
10     0.70
0      0.77
17     0.88
16     0.99
9      0.76
8      0.87
12     1.01
11     0.74
5      0.87
Name: MRR (mm2/min)Zin coated wire, dtype: float64
X_test:       Ip (A)  Ton (μs)  Toff (μs)  WF (m/min)  WT (N)  SV (V)  SF (mm/min)
6      1.0     28.0       40.0        6.0        6.0    25.0      2100.0
3      1.0     25.0       40.0        5.0        8.0    20.0      2110.0
13     2.0     25.0       45.0        7.0        6.0    20.0      2090.0
2      1.0     22.0       50.0        7.0       10.0    25.0      2110.0
y_test:  6    0.85
3      1.02
13     0.80
2      0.67
Name: MRR (mm2/min)Zin coated wire, dtype: float64

LinearRegression(copy_X=True, fit_intercept=True, n_jobs=None,
         normalize=False)
```

Fig. 4.46 Instructions to split MRR_Z Dataset into 20% test and 80% training data

```
X_train:        Ip    Ton   Toff     Wp    Wf    Wt     Sv      Sf
14  2.0  23.0  50.0  11.0  2.0  8.0  25.0  100.0
7   1.0  26.0  45.0  11.0  4.0  8.0  15.0  105.0
15  2.0  26.0  40.0  12.0  4.0  8.0  25.0   95.0
4   1.0  23.0  45.0  11.0  3.0  9.0  25.0   95.0
1   1.0  20.0  45.0  12.0  3.0  8.0  20.0  100.0
10  2.0  20.0  45.0  12.0  2.0  7.0  25.0  105.0
0   1.0  20.0  40.0  11.0  2.0  7.0  15.0   95.0
17  2.0  26.0  50.0  11.0  3.0  7.0  20.0  105.0
16  2.0  26.0  45.0  13.0  2.0  9.0  15.0  100.0
9   2.0  20.0  40.0  11.0  4.0  9.0  20.0  100.0
8   1.0  26.0  50.0  12.0  2.0  9.0  20.0   95.0
12  2.0  23.0  40.0  12.0  3.0  9.0  15.0  105.0
11  2.0  20.0  50.0  13.0  3.0  8.0  15.0   95.0
5   1.0  23.0  50.0  12.0  4.0  7.0  15.0  100.0
y_train:  14    0.920
7     1.590
15    1.800
4     1.263
1     1.200
10    1.050
0     1.400
17    1.360
16    1.680
9     1.380
8     1.070
12    1.740
11    1.030
5     1.040
Name: MRR, dtype: float64
X_test:        Ip    Ton   Toff     Wp    Wf    Wt     Sv      Sf
6   1.0  26.0  40.0  13.0  3.0  7.0  25.0  100.0
3   1.0  23.0  40.0  13.0  2.0  8.0  20.0  105.0
13  2.0  23.0  45.0  13.0  4.0  7.0  20.0   95.0
2   1.0  20.0  50.0  13.0  4.0  9.0  25.0  105.0
y_test:  6     1.560
3     1.534
13    1.400
2     0.760
Name: MRR, dtype: float64

LinearRegression(copy_X=True, fit_intercept=True, n_jobs=None,
         normalize=False)
```

Fig. 4.47 Instructions to split MRR_Z_N Dataset into 20% test and 80% training data

```
X_train:        Ip    Ton   Toff    Wp    Wf    Wt     Sv      Sf
14  2.0  23.0  50.0  11.0  2.0  8.0  25.0  100.0
7   1.0  26.0  45.0  11.0  4.0  8.0  15.0  105.0
15  2.0  26.0  40.0  12.0  4.0  8.0  25.0   95.0
4   1.0  23.0  45.0  11.0  3.0  9.0  25.0   95.0
1   1.0  20.0  45.0  12.0  3.0  8.0  20.0  100.0
10  2.0  20.0  45.0  12.0  2.0  7.0  25.0  105.0
0   1.0  20.0  40.0  11.0  2.0  7.0  15.0   95.0
17  2.0  26.0  50.0  11.0  3.0  7.0  20.0  105.0
16  2.0  26.0  45.0  13.0  2.0  9.0  15.0  100.0
9   2.0  20.0  40.0  11.0  4.0  9.0  20.0  100.0
8   1.0  26.0  50.0  12.0  2.0  9.0  20.0   95.0
12  2.0  23.0  40.0  12.0  3.0  9.0  15.0  105.0
11  2.0  20.0  50.0  13.0  3.0  8.0  15.0   95.0
5   1.0  23.0  50.0  12.0  4.0  7.0  15.0  100.0
y_train:  14    2.3660
7      2.2798
15     2.4423
4      2.3218
1      2.4829
10     2.0818
0      2.1445
17     2.2223
16     2.8251
9      2.2115
8      2.4732
12     2.4889
11     2.2525
5      2.5483
Name: SR, dtype: float64
X_test:        Ip    Ton   Toff    Wp    Wf    Wt     Sv      Sf
6   1.0  26.0  40.0  13.0  3.0  7.0  25.0  100.0
3   1.0  23.0  40.0  13.0  2.0  8.0  20.0  105.0
13  2.0  23.0  45.0  13.0  4.0  7.0  20.0   95.0
2   1.0  20.0  50.0  13.0  4.0  9.0  25.0  105.0
y_test:  6     2.5124
3      2.5045
13     2.4111
2      2.4520
Name: SR, dtype: float64

LinearRegression(copy_X=True, fit_intercept=True, n_jobs=None,
         normalize=False)
```

Fig. 4.48 Instructions to split SR_Z_N Dataset into 20% test and 80% training data

```
# Get multiple predictions
y_predict = regression_model.predict(X)

# Show the predicions
y_predict[:]

df = pd.DataFrame({'Actual': Y, 'Predicted': y_predict})
df
```

Fig. 4.49 Instructions to predict by using the trained model

Table 4.9 Results of MRR_Predicted data

MRR	MRR_Predicted
1.5438	1.510986
2.3063	2.027501
1.3125	1.084015
1.7938	1.50912
2.0625	2.065635
2.1500	2.0322
1.9000	1.8473
2.1188	2.303769
2.1250	2.6603
7.3500	8.0249
8.3167	8.881406
8.7968	9.037921
8.1917	8.263026
9.2250	9.6195
9.4250	9.8761
7.6580	8.5012
9.1750	9.2577
9.5080	9.7142
14.2417	15.578797
17.1000	16.9353
17.5167	17.291827
15.8417	15.816931
18.8670	17.1734
20.4670	19.5300
14.5875	15.055066
16.9875	17.411581
19.3500	18.7681

Table 4.10 Results of OC_Predicted data

OC	OC_Predicted
0.012	0.030651
0.018	0.039301
0.016	0.035952
0.086	0.074645
0.056	0.043296
0.068	0.051946
0.096	0.088639
0.086	0.08029
0.096	0.09594
0.106	0.09926
0.130	0.126
0.048	0.034561
0.082	0.071254
0.144	0.129904
0.084	0.068555
0.096	0.075248
0.195	0.183898
0.182	0.172549
0.181	0.163869
0.174	0.172519
0.158	0.161169
0.189	0.187863
0.225	0.216513
0.200	0.205
0.201	0.211857
0.204	0.220507
0.226	0.239158

Table 4.11 Results of SR_Predicted data

SR	SR_PredicteD
5.267	5.07827
5.330	5.223
4.130	4.367
4.400	4.609
4.530	4.754
4.130	4.598
4.670	4.741
4.600	4.685

(continued)

Table 4.11 (continued)

SR	SR_PredicteD
4.067	4.429565
6.000	6.133
6.530	6.277
6.200	6.421
7.330	7.664
7.200	7.408
7.867	8.052634
6.867	7.194893
7.800	7.339
7.930	7.484
6.600	7.087
6.930	7.331
7.530	8.076
8.867	8.717962
10.000	9.862
10.130	9.607
10.000	9.449
9.800	9.394
10.000	9.838

Table 4.12 Results of MRR_B_Predicted data

MRR _B	MRR_B_Predicted
0.58	0.615335
0.59	0.590168
0.53	0.565
0.62	0.743571
0.6	0.597933
0.64	0.616937
0.63	0.716753
0.75	0.761147
0.69	0.668479
0.57	0.634069
0.69	0.636483
0.59	0.560406
0.83	0.777716
0.82	0.606688
0.59	0.634491
0.72	0.687554
0.77	0.784919
0.69	0.734361

Table 4.13 Results of MRR_B_N_Predicted data

MRR	MRR_B_N_Predicted
1.32	1.32
1.09	1.113618
0.72	0.907236
1.46	1.858665
1.12	1.053097
0.93	0.868518
1.36	1.93494
1.29	1.242383
0.89	1.042383
1.33	1.343271
1.2	1.209172
0.93	0.958692
1.65	1.729272
1.3	1.431682
0.89	0.789606
1.75	1.767877
1.82	1.680888
0.85	0.941222

Table 4.14 Results of SR_B_N_Predicted data

SR	SR_B_N_Predicted
2.2566	2.2566
2.4543	2.419348
1.8824	2.082096
2.2927	2.578325
2.1456	2.165073
2.4008	2.454462
2.338	2.278259
2.3882	2.343258
2.2094	2.316158
1.9849	2.158614
2.2858	2.415785
2.4696	2.420904
2.4876	2.440021
2.5341	2.571159
2.3896	2.20394
2.6316	2.567204
2.4514	2.518356
2.3337	2.389375

Table 4.15 Results of MRR_Z_Predicted data

MRR _Z	MRR_Z_Predicted
0.77	0.769161
0.69	0.751902
0.67	0.734643
1.02	0.894643
0.88	0.802132
0.87	0.832121
0.85	0.887403
0.99	1.007132
0.87	0.907551
0.76	0.801077
0.7	0.687008
0.74	0.731485
1.01	0.96053
0.8	0.778279
0.75	0.75395
0.87	0.882543
0.99	0.995203
0.88	0.888203

Table 4.16 Results of MRR_Z_N_Predicted data

MRR_Z_N	MRR_Z_N_Predicted
1.4	1.4
1.2	1.109502
0.76	0.819003
1.534	1.625789
1.263	1.175924
1.04	1.125511
1.56	1.773881
1.59	1.584407
1.07	1.167657
1.38	1.447742
1.05	1.104409
1.03	0.980578
1.74	1.74515
1.4	1.434346
0.92	0.951953
1.8	1.806445
1.68	1.650085
1.36	1.273638

Table 4.17 Results of SR_Z_N_Predicted data

SR_Z_N	SR_Z_N_Predicted
2.1445	2.1445
2.4829	2.36095
2.452	2.577399
2.5045	2.633999
2.3218	2.346476
2.5483	2.381122
2.5124	2.622893
2.2798	2.423576
2.4732	2.593876
2.2115	2.202617
2.0818	2.215865
2.2525	2.407563
2.4889	2.477543
2.4111	2.423983
2.366	2.203268
2.4423	2.437036
2.8251	2.695541
2.2223	2.250966

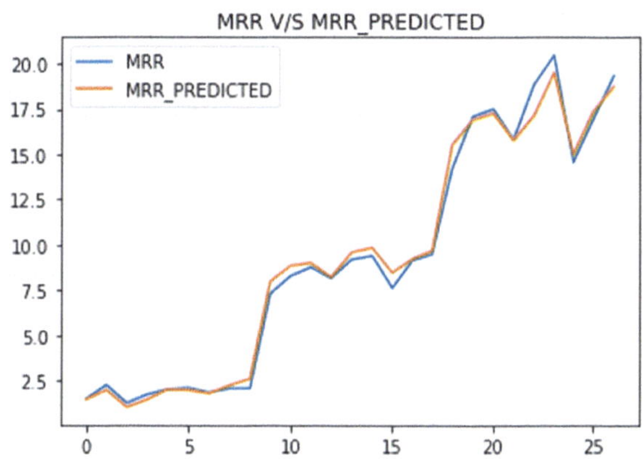

Fig. 4.50 MRR versus MRR_Predicted

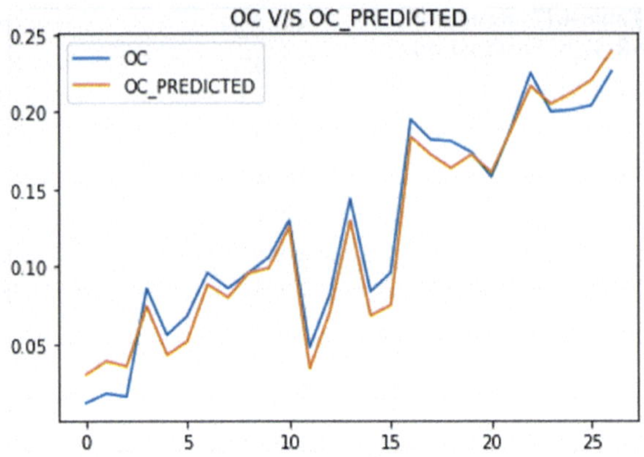

Fig. 4.51 OC versus OC_Predicted

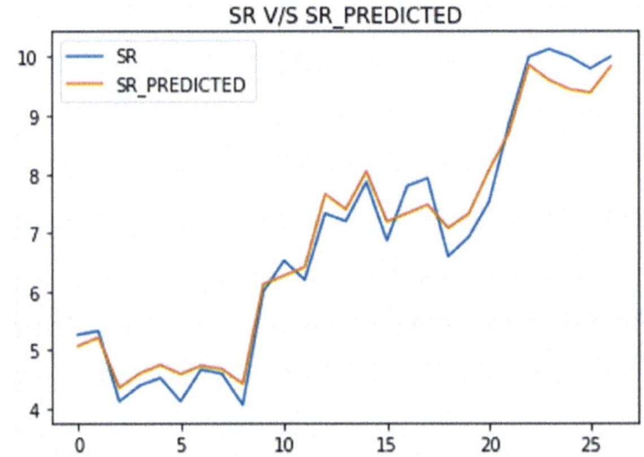

Fig. 4.52 SR versus SR_Predicted

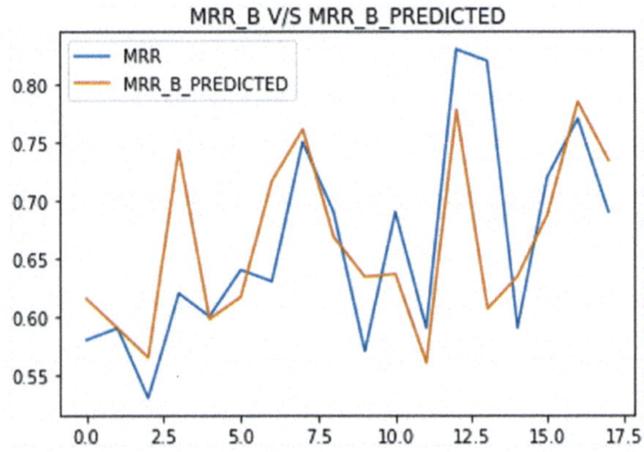

Fig. 4.53 MRR_B versus MRR_B_Predicted

Fig. 4.54 MRR_B_N versus MRR_B_N_Predicted

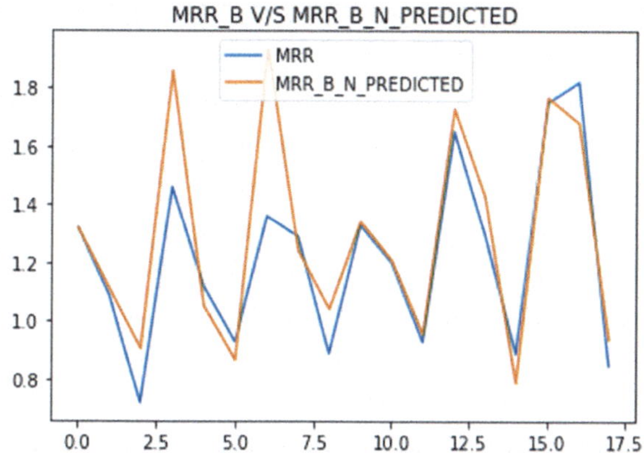

Fig. 4.55 SR_B_N versus SR_B_N_Predicted

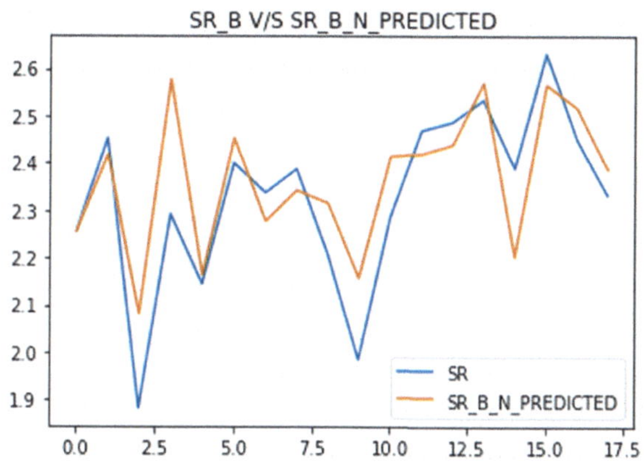

Fig. 4.56 MRR versus MRR_Z_Predicted

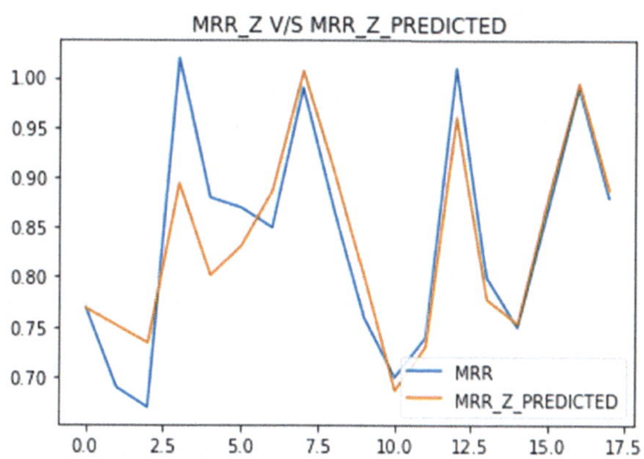

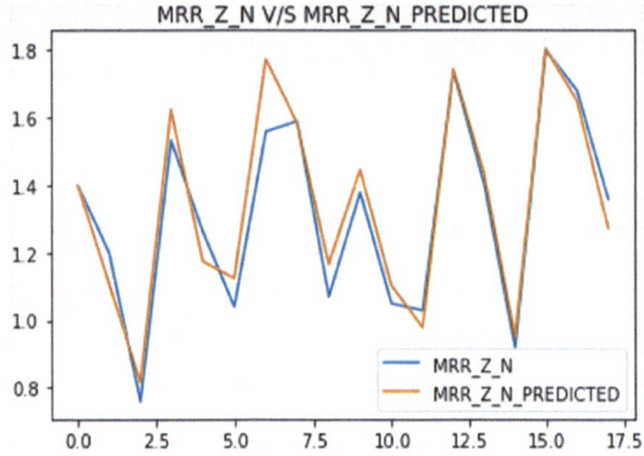

Fig. 4.57 MRR_Z_N versus MRR_Z_N_Predicted

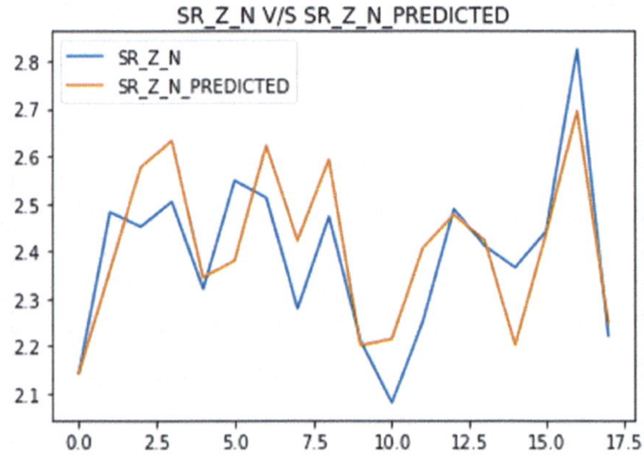

Fig. 4.58 SR_Z_N versus SR_Z_N_Predicted

References

1. Ho KH, Newman ST, Rahimifard S, Allen RD (2004) State of the art in wire electrical discharge machining (WEDM). Int J Mach Tools Manuf 44(12–13):1247–1259
2. Kozak J, Rajurkar KP, Chandarana N (2004) Machining of low electrical conductive materials by wire electrical discharge machining (WEDM). J Mater Process Technol 149(1–3):266–271
3. Lok YK, Lee TC (1997) Processing of advanced ceramics using the Wire-Cut EDM process. J Mater Process Technol 63(1–3):839–843
4. Saha N, Chakraborty S, Dey PP, Das PK (2014) Machining of ZrB_2-SiC composites by Wire-EDM technique. Trans Indian Ceram Soc 73(2):94–97
5. Tosun N, Cogun C, Tosun G (2004) A study on kerf and material removal rate in wire electrical discharge machining based on Taguchi method. J Mater Process Technol 152 (3):316–322
6. Harshdeep, Ishu Monga (2015) Study of various performance parameters of wire electrical discharge machining for H11 using Taguchi L16 array. Int J Res Appl Sci Eng Technol (IJRASET) 3 (III):884–888

7. Nagaraja R, Chandrasekaran K, Shenbhgaraj S, International Journal Of Engineering Sciences & Research Technology Optimization Of Parameter For Metal Matrix Composite In Wire EDM. Pulse 1(2):3
8. Sreenivasa Rao M, Venkaiah N (2018) Experimental investigations on surface integrity issues of Inconel-690 during wire-cut electrical discharge machining process. Proc Inst Mech Eng Part B J Eng Manuf 232(4):731–741
9. Dewangan S, Gangopadhyay S, Biswas CK (2015) Multi-response optimization of surface integrity characteristics of EDM process using grey-fuzzy logic-based hybrid approach. Eng Sci Technol Int J 18(3):361–368
10. Azhiri RB, Teimouri R, Baboly MG, Leseman Z (2014) Application of Taguchi, ANFIS and grey relational analysis for studying, modeling and optimization of wire EDM process while using gaseous media. Int J Adv Manuf Technol 71(1–4):279–295
11. Rao PS, Ramji K, Satyanarayana B (2014) Experimental investigation and optimization of wire EDM parameters for surface roughness, MRR and white layer in machining of aluminium alloy. Proc Mater Sci 5:2197–2206
12. Saedon JB, Jaafar N, Yahaya MA, Saad N, Kasim MS (2014) Multi-objective optimization of titanium alloy through orthogonal array and grey relational analysis in WEDM. Proc Technol 15:832–840
13. Goswami A, Kumar J (2014) Investigation of surface integrity, material removal rate and wire wear ratio for WEDM of Nimonic 80A alloy using GRA and Taguchi method. Eng Sci Technol Int J 17(4):173–184
14. Karim Baig MD, Venkaiah N (2014) Parametric optimization of WEDM for HastelloyC276, using GRA method. Int J Eng Dev Res 1(2):1–7
15. Sharma N, Khanna R, Gupta R (2013) Multi quality characteristics of WEDM process parameters with RSM. Proc Eng 64:710–719
16. Nourbakhsh F, Rajurkar KP, Malshe AP, Cao J (2013) Wire electro-discharge machining of titanium alloy. Proc Cirp 5:13–18
17. Shandilya P, Jain PK, Jain NK (2013) RSM and ANN modeling approaches for predicting average cutting speed during WEDM of SiCp/6061 Al MMC. Proc Eng 64:767–774
18. Shayan AV, Afza RA, Teimouri R (2013) Parametric study along with selection of optimal solutions in dry wire cut machining of cemented tungsten carbide (WC-Co). J Manuf Process 15(4):644–658
19. Kumar K, Agarwal S (2012) Multi-objective parametric optimization on machining with wire electric discharge machining. Int J Adv Manuf Technol 62(5–8):617–633
20. Yang RT, Tzeng CJ, Yang YK, Hsieh MH (2012) Optimization of wire electrical discharge machining process parameters for cutting tungsten. Int J Adv Manuf Technol 60(1–4):135–147
21. Chinnadurai T, Vendan SA (2015) Contemplating the performance measures of wire cut EDM based on process parameters for AISI 4140. Mater Today Proc 2(4–5):1067–1073
22. Chinnadurai T, Vendan SA (2016) Prediction of material removal rate and surface roughness for wire electrical discharge machining of nickel using response surface methodology. Revista de Metalurgia 52(4)

Appendix

```
import numpy as np
import pandas as pd
import seaborn as sns
from scipy import stats
import matplotlib.pyplot as plt
import statsmodels.api as sm
from statsmodels.stats import diagnostic as diag
from statsmodels.stats.outliers_influence import variance_inflation_factor
from sklearn.linear_model import LinearRegression
from sklearn.model_selection import train_test_split
from sklearn.metrics import mean_squared_error, r2_score, mean_absolute_error
%matplotlib inline

# load the dataset and replace the '..' with nan
ts_df = pd.read_excel('DATASET')
ts_df = ts_df.astype(float)
display(ts_df)

# Instruction to calculate correlation matrix
corr = ts_df.corr()

# Instruction to display the correlation matrix
display(corr)

# Instruction to plot the correlation heatmap
sns.heatmap(corr, xticklabels=corr.columns, yticklabels=corr.columns, cmap='RdBu')

# Instruction to define the plot
pd.plotting.scatter_matrix(ts_df, alpha = 1, figsize = (30, 20))
```


S. A. Vendan et al., *Welding and Cutting Case Studies with Supervised Machine Learning*, Engineering Applications of Computational Methods 1,
https://doi.org/10.1007/978-981-13-9382-2

 # Instruction to display the plot
```
plt.show()
```

 # Instruction to fetch the summary
```
desc_df = ts_df.describe()
```

 # Instructions to add the standard deviation metric
```
desc_df.loc['+3_std'] = desc_df.loc['mean'] + (desc_df.loc['std'] * 3)

desc_df.loc['-3_std'] = desc_df.loc['mean'] - (desc_df.loc['std'] * 3)
```

 # Instruction to display it
```
desc_df
```

 # Instruction to filter the data frame to remove the values exceeding 3 standard deviations
```
econ_remove_df = ts_df[(np.abs(stats.zscore(ts_df)) < 3).all(axis=1)]
```

 # Instruction to know about the removed rows
```
ts_df.index.difference(econ_remove_df.index)
```

 # Instructions to define our input variable (X) & output variable (Y)
```
X = ts_df.iloc[:, :-1]
Y = ts_df.iloc[:, 8]
```

 # Instructions to the dataset into Train and Test data
```
X_train, X_test, y_train, y_test = train_test_split(X, Y, test_size=0.20,
random_state=1)
print ("X_train: ', X_train)
print ("y_train: ', y_train)
print("X_test: ', X_test)
print ("y_test: ', y_test)
```

 # Instruction to create the object for Linear Regression model
```
regression_model = LinearRegression()
```

 # Instruction to pass through the X_train & y_train data set
```
regression_model.fit(X_train, y_train)
```

 # Instructions to get multiple predictions
```
ts_df = pd.read_excel('DATASET')
X = ts_df.iloc[:, :-1]
Y = ts_df.iloc[:, 8]
y_predict = regression_model.predict(X)
```

 # Instructions to show the first 5 predictions
```
y_predict[:]
print ("y_predict: ', y_predict)
df = pd.DataFrame({'Actual': Y, 'Predicted': y_predict})
Df
```

```
# Instructions to plot a graph between actual and predicted values
ts_df = pd.read_excel('DATASET')
X = ts_df.iloc[:, 8]
Y = ts_df.iloc[:, 9]

plt.title('ACTUAL DATASET V/S PREDICTED DATASET')
plt.plot(X, label='ACTUAL DATASET')
plt.plot(Y, label='PREDICTED DATASET')
plt.legend()
plt.show()
```

The manufacturer's authorised representative in the EU is Springer Nature Customer Service Centre GmbH, Europaplatz 3, 69115 Heidelberg, Germany. If you have any concerns regarding our products, please contact ProductSafety@springernature.com

Printed and bound by CPI Group (UK) Ltd, Croydon, CR0 4YY
24/07/2026
02175971-0003